The Complete

IRRIGATION

WORKBOOK

DESIGN, INSTALLATION,
MAINTENANCE,
AND WATER MANAGEMENT

By Larry Keesen

The Complete

IRRIGATION

WORKBOOK

DESIGN, INSTALLATION, MAINTENANCE, AND WATER MANAGEMENT

By Larry Keesen

Cindy Code, Editor

Brian Vinchesi, Technical Review

Editor: Cindy Code
Cover Design: Charlotte Turcotte
Interior Design, Layout and Production: Helen Duerr O'Halloran
Desktop Coordinator: Christopher W. Foster
Production Manager: Jami Childs
Books Manager: Fran Franzak

If you have any questions or comments concerning this book, please write:
  Franzak & Foster
  G.I.E. Inc., Publishers
  4012 Bridge Ave.
  Cleveland, Ohio 44113

Library of Congress Catalog Card Number: 95-079650

ISBN 1-883751-01-2 softcover

TABLE OF CONTENTS

design issues and describe workable solutions to design problems, determining the requirements for a project and equipment selection.

Troubleshooting irrigation system problems and deciding whether to repair or replace equipment are reviewed in this chapter.

ABOUT THE AUTHOR

Larry Keesen started worrying about water at an early age. He was 13 when he began hand watering lawns in a Denver country club as an employee of his family's landscape business. It's been uphill ever since.

Keesen has been involved in every area of the irrigation business including contracting, design, installation, maintenance and water management. He was one of the original members of Xeriscape Colorado! and served for many years on the Irrigation Association's Certification Board of Governors. He has been a consistent force in Denver's green industry through participation in the Associated Landscape Contractors of Colorado. He is also a recipient of the prestigious Bob Cannon Award for outstanding service to the industry.

Keesen Water Management, Larry's design and consulting firm, is proud of its irrigation design and evaluation work including: West Point Military Academy, Colorado's State Capitol grounds, The Tennessee Valley Authority, Kaiser Permanente and Elitch Gardens, a local amusement park. In addition to running a growing business, Keesen enjoys traveling and presenting irrigation seminars throughout the country as well as writing a wide range of articles for industry publications.

Larry is married with two children, Kristin and John, both of whom live in Denver.

DEDICATION

This book is dedicated to Mary S. Stuart, R.N., M.S., etc. Describing Mary is a book in itself, as those who know her would definitely agree. She is a remarkable individual with many accomplishments. Mary is witty, loving, intelligent, outspoken and happy. Mary has a special way of nurturing and helping people improve their lives. She is a "born" nurse, always caring for others, giving gifts and giving of her time. A successful psychotherapist and author of two self-help books, Mary is working on a third book, a biography. When Mary came into my life, I lacked confidence and esteem, struggled to write a report and was scared of public speaking. Now, I can write and thoroughly enjoy presenting seminars. Mary's influence and "therapy" over the last 15 years has taught me much. Thank you for being my best friend, companion, mentor and wife.

PREFACE

For the things I learned about life, I thank my parents Lawrence and Wilma. To my friend and partner Eric Schmidt, thanks for putting up with me the past 12 years. Eric developed the detail drawings found in the appendix of this workbook. Thanks to my two children, John and Kristin Keesen, for making my life infinitely richer with your love and support. John developed all of the charts in the appendix. Thanks to many friends in the irrigation and landscape industry who taught me much. To Cindy Code, who got me started with this project and edited the book, thank you for caring. Thanks to my friends in the Aspen Writer's group for gentle critiquing, Brian Vinchesi for technical review and Glenn Tribe for reviewing the pump chapter.

INTRODUCTION

The Complete Irrigation Workbook covers everything related to landscape irrigation systems — design, installation, inspection, maintenance, evaluation, problem solving and irrigation scheduling.

Over the past 35 years, I have obtained every book available about landscape irrigation (sprinkler) systems. Most were heavy in theory and formulae but light in practical experience including a lack of attention to the result or outcome of a poor design and installation. Little has been written about the actual experience of installing, maintaining and scheduling landscape irrigation systems.

I decided to write this book to fill in the gaps with both my experiences and those of countless others who have shared their mistakes and successes. I want to promote lower maintenance costs and higher system efficiencies. Water conservation and Xeriscape™ principles are the foundation for *The Complete Irrigation Workbook*. The high cost and limited availability of water is the driving force behind conservation. Without water our industry will vanish like the Mayan civilizations.

In this book, I describe my experiences and methods for evaluating and analyzing existing irrigation systems. An evaluation identifies system problems and recommends corrective measures necessary to improve performance. It's almost like having your automobile engine tuned up for better performance and emissions. If any irrigation system is not performing correctly it may show up as an area of brown, dry grass, but in many cases the problem will not be apparent.

The evaluation process has exposed one of the biggest problems with existing systems — operating pressure that is either too high or too low resulting in poor uniformity of coverage and damage to the equipment. An entire chapter is devoted to analyzing and correcting pressure problems.

Few contractors are aware of irrigation system safety and liability issues that can affect a businesses' bottom line, so a chapter was added with the hope that others could avoid these pitalls. Other chapters cover bidding and estimating, equipment selection, filtration and system capacity, pumps, sensors, diagnosing problems, installation techniques, preventive maintenance, system design, drip and electricity.

The *Complete Irrigation Workbook* is a guide for designers, installers, maintainers and operators alike.

1

MANAGING WATER IN THE LANDSCAPE

BEFORE DESIGNING and installing an irrigation system soil, plant and climatic conditions must first be investigated. The purpose of this chapter is to learn how to identify soil types, look at plant water use and show the effects plant materials have on irrigation system operations. The selection of sprinkler heads and nozzles, as well as the spacing and size of emitters and bubblers will also be explained.

Why is the soil type so important to the design and operation of the irrigation system and, in general, to good water management practices? Because soil is the water storage reservoir or "fuel tank" for most plant materials. Plants require water to maintain their structure, and to cool leaf surfaces by means of transpiration. Additionally, water movement into the soil is important in order to prevent runoff and promote water conservation.

The interaction among plants, the soil and available water supplies is critical to understanding how water management practices can be achieved. Soil types and textures are determined by the particle size of minerals that make up the soil, and range from very coarse sand with a particle size of up to 20 mm to clay soil with particle sizes below 0.002 mm.

For our purposes the soil categories are grouped into three types:

Soil Type	Texture	Appearance/Dry	Appearance/Wet
Sandy soils	Coarse	Loose, single-grained; flows through fingers and non-moldable when squeezed.	Compressed in the hand, it may form a ball, but will crumble easily.
Loamy soils	Medium	Slightly moldable, compressed its form is fragile when touched.	Moldable, compressed its form can be handled without breaking.
Clayey soils	Fine	Forms hard clods; surface appears cracked with loose crumbs on surface.	Very moldable and generally sticky.

Soil structure is determined by the organization and grouping of soil particles. Crumb and granular structures are the most beneficial for water storage and infiltration, but are generally only achieved after soil preparation takes place.

The irrigation system was blamed, but this soil is gravel and reflects the effects of little or no water holding capacity.

Soil pore space is the area between particles that is available to hold water and air. Sandy soil has the least amount of pore space at approximately 35 percent of total volume, while clay soils have up to 60 percent pore space. A high percentage of pore space results in a greater soil water holding capacity. When compacted, clay soils have reduced pore space that can be as low as 25 percent of the total volume. Tilling the soil prior to planting is important to create space for needed air and water.

The water holding capacity or field capacity of a soil type is defined as the amount of water that remains in the soil after the gravitational water has drained away, and after the rate of downward movement has decreased to a significant extent.

Water holding capacities of soil will vary by soil type and are generally measured in inches of water per inch of soil thickness as shown in the table below:

Soil Type	Inches of water per inch of soil
Sandy soils	0.083
Loamy soils	0.125
Clayey soils	0.167

A 12-inch layer of sandy soil holds 0.996 of an inch of water (12 x 0.083 = 0.996), and a 6-inch layer of clayey soil holds 1.002 inches of water (6 x 0.167 = 1.002). If the plant root zone is 4 inches deep in a loamy soil the water holding capacity is 0.50 of an inch (4 x 0.125 = 0.50), and a clay soil 4 inches deep has a water holding capacity of 0.668-inches (4 x 0.167 = 0.668).

WATER AVAILABILITY. Wilt results when plants can't retrieve enough moisture from the soil. The difference between field capacity and the wilting point is known as available water holding capacity (AWHC), and is the percentage of moisture available to the plant at any given time from the soil.

Management allowable depletion (MAD) is the percentage or fraction of water — of the total available water — which should be removed from the soil by the plant before irrigation occurs. In turfgrass, the MAD is commonly calculated at 50 percent of the AWHC. This is

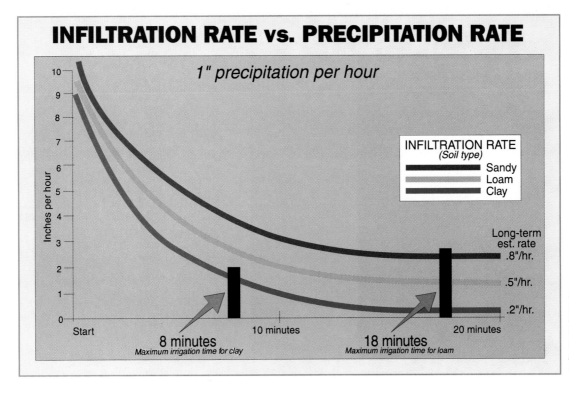

INFILTRATION RATE vs. PRECIPITATION RATE

1" precipitation per hour

Inches per hour

INFILTRATION RATE
(Soil type)
—— Sandy
—— Loam
—— Clay

Long-term
est. rate

.8"/hr.

.5"/hr.

.2"/hr.

Start

8 minutes
Maximum irrigation time for clay

10 minutes

18 minutes
Maximum irrigation time for loam

20 minutes

Ideal irrigation run times can be determined by comparing infiltration and precipitation rates.

generally a safe percentage to use to calculate frequency and quantity of water applications. The calculations can be fine-tuned with experience. For example, a clay soil with a 12-inch depth has an AWHC of 2 inches (12 x 0.167 = 2) and a MAD of 50 percent, allowing 1 inch to be used by the plant prior to scheduling the next irrigation. MAD, coupled with evapotranspiration rates (ET) indicate when to irrigate (frequency) and how much water to apply.

Saturated soil produces a steady stream of runoff and erosion, and yes, that's 110-volt wiring on the soil surface.

Soil water extraction by plants varies, but a safe assumption for calculating soil water extraction is the 40-30-20-10 rule. This rule states that 40 percent of the water will be withdrawn from the top 25 percent of the root zone, 30 percent will be withdrawn from the second 25 percent of the root zone, 20 percent from the third 25 percent and 10 percent from the lowest 25 percent of the root zone.

Transpiration occurs when the plant pulls water through the roots and stomates and discharges it through the leaf into the air as a vapor. This reaction results from a difference in tension or atmospheric suction. The plant uses this water to draw up nutrients, maintain structure and to cool the leaf surface.

ET is the combination of transpiration and water that is evaporated from the soil surface. It is usually defined as the quantity of water in inches or millimeters that needs to be replaced in the soil to maintain ideal growth and appearance. Evaporation from the soil surface usually affects only the top few inches of soil. Daily irrigation of a turfgrass based in a clay soil can result in a higher soil evaporation rate than watering every other day or every third day. This is not much of a problem in sandy soils because the water drains faster. Crop cover and mulches also reduce the amount of soil evaporation.

Several climatic factors affect ET including air temperature, wind speed, relative humidity and solar radiation. Solar radiation — the primary climatic factor — is the source of energy necessary to convert water into vapor in both plants (transpiration) and soil (evaporation).

Solar radiation peaks in late June in the Northern Hemisphere and December in the Southern Hemisphere at latitudes above 22 degrees. Temperatures usually peak in June or July in the North, and December and January in the South. After these months, the ET rate begins to decline leading to reduced irrigation through the remainder of the season.

Soil Type	Basic intake rate	Turf/soil intake rate
Sandy soils	0.60 to 1 in/hr	0.30 to 0.50 in/hr
Loamy soils	0.30 to 0.60 in/hr	0.15 to 0.30 in/hr
Clayey soils	0.10 to 0.30 in/hr	0.05 to 0.15 in/hr

Weather stations can provide ET data on a daily basis to determine the next irrigation cycle. For instance, California and Arizona have a series of weather stations in various state regions that provide a daily ET rate for irrigators. Historical monthly ET rates (not as accurate as current data) can also be used for scheduling irrigation. These are available from the National Climatic Center in Asheville, N.C., or from your local weather service.

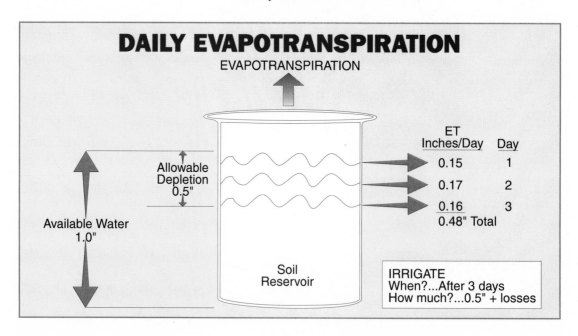

DAILY EVAPOTRANSPIRATION

EVAPOTRANSPIRATION

	ET Inches/Day	Day
	0.15	1
	0.17	2
	0.16	3
	0.48" Total	

Allowable Depletion 0.5"

Available Water 1.0"

Soil Reservoir

IRRIGATE
When?...After 3 days
How much?...0.5" + losses

The soil infiltration rate or intake rate is the maximum rate at which the water enters *bare soil* during irrigation, and is usually measured in inches per hour (in/hr). Water moves into the soil first by gravity and then by both gravity and capillary action. The top layer of soil fills with water to field capacity before the water moves downward. Each subsequent layer does the same. Soils high in organic content will generally have a higher infiltration rate than soils low in organic content.

Turf cover as well as compaction, slope and thatch will decrease the infiltration rate. During the first few minutes of irrigation the infiltration rate is very high at 5 to 10 inches per hour and then rapidly decreases to the rates described above.

Sprinkler heads with a precipitation rate of 1-inch per hour have a maximum irrigation run time in bare clay soil of eight minutes. On the other hand, 18 minutes of run time is acceptable for loamy soil before runoff occurs. With a *turf crop cover,* a maximum run time of *five minutes for clay soil* and *10 minutes for loamy soil* is recommended. Repeat the cycle after 30 minutes if additional water is needed.

Plant roots need oxygen to function and grow. A soil at field capacity most of the time is not conducive to healthy plant growth because oxygen movement in the soil is limited. This can be a problem in clay soils but can be avoided by infrequent or "just in time" irrigation. In sandy soils this does not present much of a problem. Frequent aeration allows oxygen to readily enter the soil, improving the intake rate of the soil. The depth of the active root zone for many species of turfgrass is less than 6 inches, and in some cases may be only 1 to 2 inches. Roots do not have the ability to search or hunt for water, and will not grow through dry soil in search of water.

When selecting irrigation equipment and nozzle sizes it's helpful if the application rate of the sprinkler head matches the intake rate of the crop cover/soil combination. With certain soil types and sprinkler head combinations this is not possible. For instance, a clay soil combined with pop-up spray heads is not recommended. Clay soils have a very slow infiltration rate at less than 0.15 inches per hour, while the pop-up spray head has an average application rate or precipitation rate of 1.5 inches per hour. Bubbler heads can be spaced further apart in a clay soil because the capillary action of the clay soil increases the lateral movement of water, whereas there is little lateral movement in a sandy soil.

The information in this chapter is critical for understanding how water moves in and out of soils, how water is stored, how plants use water and how to practice good water management. The next chapter continues with water management including irrigation scheduling, calculating precipitation rates, controllers and programming and weekly applications.

QUESTIONS:

1. What type of soil has the least amount of pore space?

_____

2. What is the approximate infiltration rate for a bluegrass lawn with a clay soil?

_____

3. What is the available water holding capacity (AWHC) of a 6-inch layer of loam soil?

_____

4. What amount of water can be used by a plant prior to scheduling the next irrigation in a fescue turfgrass, growing on sandy soil with a 4-inch active root zone?

_____

5. With a daily ET rate of 0.20 inches, what is the irrigation frequency and amount of water applied for a turfgrass lawn on clay soil with a 5-inch active root zone?

_____

6. What is the maximum run-time for small spray heads watering turfgrass on a clay soil?

_____

7. When does the solar radiation peak?

_____

8. What can be done to prevent plant oxygen starvation?

_____

9. What factors affect water infiltration into the soil?

_____

10. What increases soil water capacity and infiltration?

_____

(Answers to these questions are found on page 251.)

2

SCHEDULING FOR OPTIMUM PLANT GROWTH

THE FIRST SIGN of trouble was apparent when we heard the squish, squish, squish of our feet as we walked the boundaries through the wet turf at Sunbird Town homes. During the summer of 1988, we were asked to assess the high water bills and evaluate the irrigation system for the five-and-a-half acres of turf and landscaping surrounding 250 two-story town houses located in Denver, Colo.

The purpose of an irrigation system evaluation is to conserve water, reduce water costs and promote healthier plant growth. Malfunctioning irrigation systems can overwater, leak, cause damage to structures and create the potential for injury.

A sprinkler system evaluation identifies system problems and aides in making recommendations for corrective measures needed to improve performance. It's almost like having your automobile engine tuned up for better performance and emissions. If any irrigation system is not performing correctly it may show up as an area of brown, dry grass, but in many cases the problem will not be readily apparent.

When evaluating the system at Sunbird, we discovered one problem in the way the irrigation schedule was set. More than 120 inches of water were being used annually at the site whereas average ET rates for Denver turfgrass with a system efficiency of 80 percent should never be above 30 inches, and can be as low as 12 inches or less with proper management.

When the system was correctly scheduled in 1989, Sunbird saved more than $11,000 annually in water costs. Additionally, water use was reduced by 35 percent. That's enough water (75 million gallons) to supply 100 families with average water usage for more than 12 years. This story is too familiar on many landscaped sites across the United States. Generally, auditing and evaluation shows that overwatering occurs on most properties around the coun-

try. When you think of the huge waste of water —
a precious resource — and the potential environmental damage to structures and pavement from
irrigation system runoff, it makes you shudder.

When contractors install a new irrigation system,
the initial irrigation schedule is set for total saturation or
the "rice paddy effect." Additionally, many contractors and designers fail to provide the owner with
proper scheduling guidelines and techniques. This
can be a huge liability for any designer, installation
contractor and/or maintenance contractor, but
much of this risk can be prevented by properly
scheduling the irrigation system.

In the preceding chapter we learned about soils,
plant water use and irrigation system performance. The
purpose of this chapter is to learn proper irrigation scheduling techniques and the equipment required to
attain this proper landscape watering.

A system evaluation at Sunbird identified the problems and provided cost effective solutions.

GETTING STARTED. The first step in scheduling is to
determine the application rate or the precipitation rate
of the sprinkler heads for each zone throughout the
irrigation system. This can be done using a variety of
methods:

1. Using a formula that will calculate the theoretical application rate for any
   sprinkler head. This is the most common method and is generally used when
   designing systems.

2. Place catchment devices in the landscape areas to measure the amount of
   water that is applied for a given time period in millimeters and/or inches. This
   rate of application can be used to calculate accurate irrigation schedules
   and will be discussed in later chapters.

3. Flow meters can be used to calculate application rates if the size of the area
   served by a zone is known.

4. Another method is to measure the operating water pressure and nozzle diameter at the head and determine the flow rate using a nozzle size/pressure chart
   (see appendix) for the following formula.

The common formula for determining the application rate in inches per hour is:

$$\frac{96.3 \times GPM}{SPACING \times SPACING \ (AREA)} = \text{Application rate in inches per hour}$$

The formula for the constant 96.3 converts cubic inches of water to inches per square foot per hour as follows:

$$\text{One gallon} \quad = 231 \text{ cubic inches}$$
$$\text{One square foot} = 144 \text{ square inches}$$

$$\frac{231}{144} = 1.604 \text{ (in./sq. ft./gal./min.)} \times 60 \text{ min.} = 96.3$$

Using manufacturers' catalogs, you can determine the performance data for a pop-up spray head with a full-circle nozzle. The result is a 5-foot radius of coverage with a flow rate of 4.0 gallons per minute (GPM) at 30 pounds per square inch (PSI), and a recommended square spacing of 15 feet. The formula looks like this:

$$\frac{96.3 \times 4.0 \text{ (GPM)} = 385.2}{15 \text{ ft.} \times 15 \text{ ft.} = 225} = 1.71 \text{ inches per hour}$$

If the spacing is triangular, the bottom part of the formula represents the spacing (15 feet) multiplied by the distance between rows (or the height of the triangle). To determine the height of an equilateral triangle (all sides of the triangle are the same length), take the base (15 feet) times 0.866 to determine the distance between the rows.

$$\frac{96.3 \times 4.0 \text{ (GPM)} = 385.2}{15 \text{ ft.} \times (15 \text{ ft.} \times 0.866) = 194.85} = 1.98 \text{ inches per hour}$$

INTAKE RATES. Now that the application rate of the zone is known, refer to the information in the previous chapter about soils and soil intake rates. Identifying the soil type and crop cover is required to determine the intake rate.

In bluegrass turf with a clayey soil and without soil preparation, the intake rate will be very low — approximately 0.05 to 0.10 inches per hour. If the sprinkler head application rate is 1.71 inches per hour and 0.50 inches needs to be applied, the run time for the zone will be about 18 minutes (0.50 in./1.71 inches = 0.292 x 60 min. = 17.52 min.), but runoff or puddling will occur after about eight minutes. To avoid runoff and promote more efficient irrigation, the runtimes should be reduced to six minutes with three start times or repeat cycles. To be safe, always use short runtimes of five to seven minutes for pop-up spray heads and repeat cycles as needed.

The next step is to determine the size and holding capacity of the "fuel

Turf and shrubs have different water requirements and should be zoned separately.

tank." This information allows the operator to ascertain the frequency between waterings. With a clayey soil, the chart in the previous chapter indicates a water holding capacity of 0.167 inches of water per inch of soil. Next, the depth of the active root zone needs to be determined using a soil probe. If it's 5 inches, multiply 5 inches times 0.167 which indicates a total water holding capacity of 0.835 inches of water.

PLANT WATER USE. Now determine the plant water use to determine when the underground fuel tank needs to be refilled. Plant water use varies widely through the seasons.

Two ways to predict plant water requirements are to use current weather station information to calculate evapotranspiration (ET) or historical ET and crop co-efficient data. Current weather station data is best because historical data is an average of up to 30 years, and quite often is inaccurate as global warming occurs. Using the historical ET data from the "Irrigation Association Landscape Irrigation Auditor" training handbook, the ET rate for Cleveland, Ohio is:

Proper irrigation scheduling saves water and reduces runoff and erosion.

JAN	FEB	MAR	APR	MAY	JUN	JUL	AUG	SEP	OCT	NOV	DEC
0.0	0.0	0.0	1.0	2.0	4.0	6.0	5.0	3.0	1.0	0.0	0.0

With an ET rate of 5 inches and a crop co-efficient of 80 percent the plant water requirement for the month is 4 inches (5 X 0.80) of water. ET is usually based on a field crop such as alfalfa, and a crop coefficient (or percentage) is used to determine the specific plant water requirement. Bluegrass turf has a crop coefficient of 80 percent of ET.

After reviewing and understanding the management allowed depletion (MAD) formula, the monthly water requirement of four needs to be converted to a daily water requirement by dividing 4 inches by 31 days. After calculations, this amounts to a daily average plant water requirement of 0.13 inches.

To determine the frequency between waterings, divide the amount of water available for plant use (WAP) — (MAD is 50 percent of 0.835, see above) or 0.42 inches of water that the plant can remove from the root zone without risk of wilting — by the adjusted daily ET rate (ET times crop co-efficient) of 0.13 which equals 3.23. Irrigation schedules will be every third day, unless rainfall occurs. The run time can then be adjusted for an application of 0.39 inches per irrigation which is 14 minutes or three, five-minute cycles. This application amount does not include corrections for efficiency and assumes no rainfall for the month. See Chapter 4 for information on efficiency.

CALCULATING IRRIGATION FREQUENCIES

F = MAD/ETA

0.42/0.13 = 3.23 or every three days.

F = Frequency between waterings

MAD = Management allowed depletion for plant use is 0.42 inches of water (clay soil holds 0.167 inches of water per inch of soil depth, times a 5-inch root zone = 0.835 inches and a MAD of 50 percent = 0.42).

ETA = evapotranspiration adjusted for turf (Sept. = 5 inches/31 days = 0.16 x 0.80 percent (crop co-efficient) = 0.13 inches

Irrigation scheduling must also take into consideration those days where watering may not be desirable, such as high usage periods during the day or week, prior to mowing, daily periods of high wind, when water is available and special events.

Take the daily ET and multiply by 7 days to determine the weekly requirement (0.13 X 7 = 0.91 inches). Then divide the answer by three watering days (0.91/3 = 0.30 inches per irrigation). Run time is determined by the irrigation water requirement divided by the application rate times 60 minutes (0.30/1.71 inches X 60 = 10.5 minutes). Two cycles at five or six minutes each works well. This does not include system efficiency.

WHEN TO WATER. The best time of day for irrigation is between the hours of 10 p.m. and 6 a.m. for the following reasons:

1. Potential soil compaction is reduced by the absence of foot traffic on wet grass.

2. The ET rate is much lower at night because it is usually cooler, there is less wind (especially in the early morning hours) and little solar radiation.

3. Less water is wasted because of wind deflection.

4. Improves public relations by reducing the visibility of water waste.

5. Avoid watering before 10 p.m. to avoid fungus problems.

6. At least a 10 percent savings in water is achievable by watering at night.

CONTROLLER FEATURES. Once an irrigation schedule is determined, look at the irrigation controller features and how they help to conserve water, lower maintenance costs and provide for a beautiful landscape. Landscape contractors, property managers and homeowners who maintain irrigation systems are looking for ways to save time, money and water. This can be accomplished by evaluating irrigation system controllers to determine if they are dependable, water conserving and, above all, user friendly.

First, a brief history of landscape irrigation controllers. The older controllers generally included a day wheel, a 24-hour dial and a station dial. These were operated by a series of clock motors and micro switches and are referred to as mechanical controllers. In the 1970s, Johns-Manville marketed the first solid-state electronic digital controller which had no moving parts except for a keyboard for data entry. Since then, many companies have developed solid-state electronic and electro-mechanical controllers along with computer software to operate irrigation systems.

Electronic controllers are more dependable than ever and offer many new features which were not previously available. Some of the best features of the newer controllers are:

1. Timing that is accurate within seconds.

2. Three or more start times allow repeat cycles to reduce runoff.

3. The selection of any combination of irrigation days is available. For example, a Monday, Wednesday, Friday schedule or every third day, every fifth day or even once per month can be programmed.

4. Electrical surge protection is built in to prevent damage and maintain backup programs for controller operations.

5. A water budget feature allows quick changes in the water application rate to adjust for ET.

6. Communication can be established between the controller and flow sensors, soil moisture sensors, rain and wind shutoff devices, weather stations and central control systems.

7. Two or more separate programs can be chosen that allow for independent irrigation schedules to accommodate different plants.

8. Start times are programmed at any minute of the day instead of on the hour or every 15 minutes.

9. Cascading start times can automatically start the second cycle after the last valve of the first cycle shuts down.

10. Programmable cycles for testing the irrigation system can also be used.

FOUR WATER CONSERVATION TECHNIQUES EVERYONE SHOULD KNOW:

- If you can see your own footprints in the grass you have just walked on, you need to water!
- Use short run times and several cycle starts per irrigation period, and water as infrequently as possible.
- Water between 10 p.m. and 6 a.m. to decrease water loss through evaporation.
- Pay attention to the seasons: Plants need little or no irrigation water through the winter and an increasing amount each month till the demand peaks in July, and begins to taper off each month as the season cools.

11. Remote control compatibility can be accessed.
(See manufacturers' product literature for complete information on controllers.)

Consider some other controller characteristics which can reduce irrigation and landscape maintenance costs.

Is the controller in good physical condition and free of rust? If the cabinet is damaged, water may enter the controller causing short circuits and corrosion. Is your controller user friendly? If maintenance personnel have continual problems understanding and operating the controller, the landscape will suffer. Get some input about the controller from the person who is operating the system.

In the event of a power failure, a built-in standby or backup program is helpful (for solid-state electronic controllers) because it will operate the system even if the custom program is erased. Most of the electro-mechanical controllers don't require a backup battery because the program is set with a series of dials instead of programming a computer chip.

When buying a controller make sure that it will meet the needs of your system(s) for at least 10 years. Those needs should include control features that promote water conservation and provide some degree of flexibility for future changes. Don't buy features which will never be used or that are non-essential to the irrigation system function.

QUESTIONS:

1. What is the best time of day to water?

2. What is the application rate for a large rotor head with a flow rate of 15 GPM at 60 PSI and a triangular spacing of 55 feet?

3. What is the application rate of a zone that has a flow rate of 50 GPM and covers 12,000 square feet?

4. How many cubic inches are there in a gallon?

5. With a monthly ET rate of 4 inches, a crop co-efficient of 85 percent, square head spacing at 65 feet and a flow rate of 19, what is the total monthly irrigation run time in minutes?

6. What is the irrigation frequency and runtime for the above scenario with a clay soil and a root zone of 6 inches?

_____

7. What does the constant 96.3 represent in the application rate formula?

_____

8. What is the crop co-efficient of bluegrass turf?

_____

9. What do footprints in the grass mean?

_____

10. How is the height of an equilateral triangle determined?

_____

(Answers to these questions are found on page 251.)

PRESSURE, THE UNSEEN MENACE

IMPROPER WATER PRESSURE, resulting in the loss of precious water supplies, is the number one problem facing the landscape irrigation industry. Pressure problems are typical of most existing systems because many designers and contractors either do not understand the importance of pressure control, or they don't care.

Recently, while evaluating an irrigation system, we found a static pressure of 62 PSI on the building hose faucet and an operating pressure of up to 72 PSI at the pop-up spray heads and small impact rotors. Since the building was protected by pressure reducing valves (PRV), we went into the mechanical room and checked the pressure upstream of the PRVs. Upon inspection, it was determined that the pressure was 113 PSI at the point of connection for the irrigation system.

Needless to say, the rotors ran like "machine guns" creating a bank of mist that drifted away from the area for which it was intended, while greatly reducing the longevity of the head. The pop-up spray heads misted and shrieked from the high flow velocity, and some leaks were apparent around the heads and wiper seals. For example, one impact rotor head zone was twice as large as the rest, and water pressure was low at 30 PSI. Improper operating (dynamic) water pressure is the most common and often unnoticed problem with landscape irrigation systems.

When water pressure at the sprinkler head is either too high or too low, both water pressure situations cause a distortion of the spray pattern resulting in poor uniformity of coverage. Consequently, when water exits the nozzle of the irrigation head it uncharacteristically explodes into droplets of water.

Low operating pressure causes the water to explode into larger droplets which produces soil compaction and also reduces the effective radius of coverage. High pressure causes the water to explode out of the nozzle into a higher number of very tiny droplets that range in size from 1 millimeter (moderate rain) to 0.10 mm (mist).

A 1 mm drop falling from 10 feet in a 3 mph wind will drift 5 feet but a 0.10 mm drop will drift 50 feet. This reduces the effective radius of coverage and causes the water to appear as a drifting mist. This mist or group of small droplets will evaporate much faster than larger droplets, and will easily drift away from the irrigated area.

Imagine the water wasted by wind drift in a 5- or 10-mile-per-hour wind. If the irrigation system is operated with improper pressure, water is wasted.

WIND DRIFT
DISTANCE DRIFTED DURING A FALL FROM 10 FEET IN A 3 mph WIND

DROP DIA. (mm)	TYPE OF DROP	DISTANCE DRIFTED
1.0	Moderate rain	5'
0.5	Light rain	8'
0.3	Drizzle	13'
0.2	Drizzle	17'
0.1	Mist	50'

Source: Delavan, 1982

For more information on droplet size, wind drift and the sail index, contact the Center for Irrigation Technology, California State University, Fresno, CA 93740-0018.

The Soil Conservation Service of the U.S. Department of Agriculture has tested sprinklers and published a recommended operating pressure range for various sprinkler flow rates. These are as follows:

SPRINKLER FLOW GPM	USDA RECOMMENDED PRESSURE RANGE PSI		WETTED DIA. (no wind)
2	20	25	75'
4	30	35	79'
6	35	40	88'
8	35	40	96'
10	40	45	100'
15	40	45	117'
20	45	50	123'
30	50	55	134'
50	60	70	175'
100	70	90	210'

Ideal operating pressure (and therefore adequate droplet size to meet the needs of plant materials) for small pop-up spray heads is 25 to 30 PSI. A pressure of 30 to 50 PSI is suggested for most rotor head applications.

Pressure problems are sometimes hard to identify visually when pressures are only moderately high — 30 percent to 60 percent higher than the manufacturers' recommended operating pressure.

Some or all of the following symptoms may be apparent when pressure problems exist:

HIGH PRESSURE	**LOW PRESSURE**
Reduced radius of coverage	Reduced radius of coverage
Floating fine mist	Large water droplets
Dry areas between head patterns	Doughnut shaped dry area
Fast rotor rotation speed	Slow rotor rotation speed
Rotor rotation failure	Rotor rotation failure
Leakage at the head	Head failure to set and seal

A word of caution: Misting can occur when heads are operating at the proper pressure if heads are tilted, are too low or if the turf interferes with the spray pattern. The unintended interruption causes the water stream to break up into smaller droplets.

To verify the operating pressure at the head, a tee with a pressure gauge under the nozzle of a pop-up spray head can be installed. Then turn on the zone and read the operating pressure. With rotor heads, insert a Pitot tube (with a pressure gauge attached) into the stream of water that exits from the nozzle. Don't insert the Pitot tube into the nozzle because this will result in a false reading. Keep the Pitot tube about 1/4-inch away from the face of the nozzle.

A PSI reading can be taken with a Pitot tube, but make sure it's placed about 1/4-inch away from the face of the nozzle and in the middle of the water stream.

Rotor head operating pressure can also be determined by installing a temporary tee with a pressure gauge under the head or removing the nozzle and inserting a pressure gauge in its place. The latter method should only be used when the flow rate of the head is less then 10 percent of the total zone flow rate, otherwise the pressure reading may increase.

A 5/32 nozzle operating at 30 PSI will have a flow rate of 4 GPM. If the pressure is increased to 50 PSI the flow will be 5.2 GPM. At a 70 PSI it will be 6 GPM. As you can see, the higher the pressure the greater the flow will be from the nozzle.

High water pressure can also cause

These rings are the result of low operating pressure. A booster pump was installed and the problem was corrected.

A pressure gauge showing 90 PSI at the nozzle, and it hasn't blown off yet. The higher the pressure the greater the flow from the nozzle.

surges in lateral lines, especially if the lateral is drained or partially drained after every cycle. This results in damage to the equipment, water leaks and a reduced system lifespan. Always install heads with check valves to prevent low head drainage (where water drains partially or completely out of the lateral line through the sprinkler head after each irrigation cycle is completed), to significantly reduce maintenance costs and to save water at the same time.

Once the pressure is measured and the problem is identified how do you make corrections? For high pressure, pressure reducing valves should be installed throughout the system to provide for optimum thrust. Replacing all pop-up spray heads with new 4-inch (or higher) pop-ups designed with an individual pressure reducing device (set for 30 PSI) and installed as an integral part of the pop-up head can solve the pressure control problem for pop-up spray heads.

Pressure reducing valves can be installed in the system with an automatic control valve to control pressure for the rotor head zones. Separate plastic pre-set PRVs can also be installed under the rotor heads. In the near future we will see rotor heads with built-in pressure reducing valves. Whenever the static pressure in a system is more than 75 PSI, install an adjustable PRV at the point of connection to protect the system from unnecessary water surges.

If the water pressure is low, try reducing the nozzle size (smaller nozzle/lower flow) if spacing is 50 percent to 55 percent of the original design radius. Another method of correcting low pressure is to split the zone and add another electric control valve. If everything else fails, a booster pump should be considered. (Some water agencies will not allow booster pumps for irrigation from potable water supplies).

Low water pressure at the head can also be caused by a restriction or blockage in the system if there is a large variation between the static pressure and the operating pressure. Check the main shutoff valve, isolation valves and flow controls on the electric valves for restrictions including blockages or partial closure. Check valves in the mainline or backflow preventer can also fail

High pressure and wind are the reason for the dry areas in this athletic field at the Air Force Academy.

and cause severe restrictions.

Several years ago, Keesen Water Management designed a streetscape irrigation system for a small residential development. After the system was installed, the contractor called and complained that the system wasn't working properly. A site visit the next morning to operate the system revealed it was only marginal. This site had approximately 50 feet of elevation change from the entry down to the bottom of the cul-de-sac. Checking the static pressure at the bottom, it was 20 PSI. Another pressure reading part way up the hill was 15 PSI. Seeing these pressure readings caused all sorts of thoughts to race through our minds. Did we miscalculate something? Was this system ever going to function properly?

Next we checked the pressure on the home across the street (a home not part of this development), and the pressure was 80 PSI. This led us to believe there was a restriction in the system and we proceeded to check every valve with the contractor. We found the valves all wide open leading us to approach a few nearby homeowners for their input. They disclosed they were having similar pressure problems.

They said the pressure was OK in the morning, but was so low in the evening that they could not use the shower. Further investigation revealed that this subdivision had a new water main extending into the site and, possibly, the isolation valve on the water main could be partially closed. This allowed the pressure to build up in the city water main overnight and create higher pressure in the morning. Subsequently, the water department found the isolation valve barely open and the situation was soon corrected.

As you can see, a lot of water can be wasted if the pressure is not correct. Are the systems that you design, install and/or maintain operating at the proper pressure? Correct irrigation system operating pressure and the control of this is a must for designers and contractors who want to provide their clients with an efficient, quality product.

If not, you may be held liable for wasting water and structural damage to asphalt surfaces; foundations, walks and patios (settlement and breaking); poor uniformity of coverage; and drainage problems. This is THE NUMBER ONE PROBLEM with most systems today. The professional irrigation industry needs to look at this and correct the pressure problem before the World runs out of water.

QUESTIONS:

1. How far will mist drift in a 3 mph wind from a height of 10 feet?

2. What is the optimum operating pressure for a large rotor head with a flow rate of 15 GPM and a triangular spacing of 55 feet?

3. What is the best operating pressure for pop-up spray heads?

_____

4. Visually, what do high pressure droplets appear as in a large group?

_____

5. How is the operating pressure obtained for rotor heads?

_____

6. Will the nozzle flow increase if the water pressure is raised from 30 PSI to 60 PSI?

_____

7. When water pressure is too high or too low, what do they manifest in common?

_____

8. Large water droplets are an indication of what?

_____

9. How can system performance be improved when the operating pressure is too low?

_____

10. Can high water pressure cause surges in lateral lines?

_____

(Answers to these questions are found on page 251.)

4

IN SEARCH OF UNIFORMITY

POOR UNIFORMITY IS the primary unseen cause of dry and wet areas, high water costs and wasted water. Irrigation system uniformity will affect maintenance costs and profitability for any site. Three townhouse associations in the Denver area have won six-figure cash settlements from contractors because of improper design and installation of irrigation systems.

The contractors all did the same thing — they stretched the spacing between the heads to 70 percent and 80 percent of the diameter of intended coverage when the manufacturers' recommended spacing was 50 percent of the diameter. The consequence was poor uniformity, which resulted in damage to turf and asphalt from the overwatering required to prevent the turf from wilting and dying.

When equal amounts of water from the irrigation system are applied to each square foot of soil surface area the system has a very high (100 percent) uniformity, resulting in less water use, fewer drainage problems and less plant disease. Uniformity measures the mechanical preference of the irrigation system. The minimum acceptable uniformity for turf in most circumstances is about 80 percent.

Uniformity for sprinkler irrigation systems using rotor heads can be 90 percent or higher, but in real life a high percentage of existing systems are below 70 percent. The average is around 50 percent. In small, irregular and narrow areas, pop-up spray heads must be used instead of rotor heads because the minimum rotor radius is about 20 feet. Pop-up spray head uniformity will not exceed 70 percent and is generally much lower than the rotor heads. Low uniformity of coverage is usually the result of an inadequate design, installation deficiencies, lack of proper maintenance or a combination thereof.

Uniformity is important for turfgrass because every square inch of area has plant roots with very shallow root zones. In fact, 75 percent of the roots are often in the top 1 inch of soil. Trees and shrubs are similar proportionately with the top 1 to 2 feet of soil containing most of the total plant root system. Hence we waste water with deep root watering at the 3- or 4-foot level

These two rows of 15-foot radius spray heads are at least 30-feet apart and require at least another row of heads to provide acceptable uniformity.

because only 10 percent to 30 percent of the roots are located here.

To assure good uniformity, the industry standard for sprinkler head spacing is 50 percent of diameter or "head to head" spacing in which the water from each head reaches the adjacent heads. Larger turf areas similar to golf courses and parks are the exceptions. The larger areas enable the relative easy use of triangular spacing with spacing of 55 percent to 60 percent. Rectangular spacing, on the other hand, could be 50 percent by 60 percent. Spacing at 50 percent of diameter will usually accommodate winds of up to 5 mph, but if wind velocity is greater than 5 miles per hour during the watering period, the spacing should be derated approximately 1 percent for every 1 mph of wind speed. A prevailing wind of 12 mph requires a spacing of 44 percent of the diameter. A head with a 30-foot diameter of coverage requires a spacing of 13.2 feet or 44 percent of the diameter.

Use low-angle nozzles where wind conditions exist. Follow manufacturers' recommended spacing and don't forget to reduce the spacing for wind conditions to avoid liability and waste.

The head spacing at this home is too close for the two heads in the center, and too far apart from the center heads to the quarter heads by the driveway.

Other factors that affect uniformity are sprinkler operating pressures, sprinkler distribution profile and slopes in which the uphill radius is reduced and the downhill radius is expanded, taking on an egg shape as the gradient increases. The standard for head spacing on slopes is to reduce the spacing horizontally (heads at the same elevation) along the slope by 1 percent for every 1 percent of slope change beyond a 20 percent grade or a 5 to 1 slope. Vertical spacing on the slope is not changed (from top to bottom), except the spacing is increased between the bottom of the slope and the first row of heads because of the increase in the downhill spray pattern, and reduced as the last row interfaces with the top of the slope. (See appendix beginning on page 167 for slope charts.)

TESTING UNIFORMITY. If you want to test the uniformity of a system as well as the net application rate, place catchment devices (conical catch devices, rain gauges or cans) in landscaped areas to measure, in inches, the amount of water that is applied for a given time

A concrete bench disturbs the spray pattern and ruins the uniformity of coverage. At least it's always clean.

period. This rate of application can then be used to calculate accurate irrigation schedules. If the catchment device is located between two zones or in a number of different zone areas, those zones must then operate for the same amount of time. Operate the system for five to 10 minutes for spray heads or 20 to 30 minutes for rotor heads, then measure the amount of water in each catchment device. You will probably see some containers with a lot more water than others.

Of the numerous catchment tests we have performed over the last 12 years we have consistently seen problems with uniformity. For instance, at a 5-acre multi-family housing site the uniformity and application rate of pop-up spray heads was measured at three locations within a single irrigation zone using the "can" test method. The results were 0.60 of an inch, 1.1 inches and 2.3 inches per hour. The range was 0.60 to 2.3 inches per hour. The area with 0.60 inches was dry and stressed receiving too little water while the area with the 2.3-inch measurement was soft and wet. As you can see,

the uniformity was very poor and the wet area was receiving four times the amount of water than the dry area, or an additional 1.7 inches of water per hour. This range results in some areas being over-watered by up to 283 percent.

This rotor is leaning to the left and the spray is barely clearing the grass, resulting in poor uniformity.

Next, the application rates of large rotor heads at three locations within a zone were measured. The results were 0.38, 0.41 and 0.44 inches per hour resulting in an additional 0.06 inch of water or a 16 percent increase. This area had a very high distribution uniformity of 93 percent.

Distribution Uniformity (DU) is one of the methods of calculating system uniformity and is measured as a percentage which indicates how closely the driest area compares to the average precipitation rate of the irrigated area. The driest area is described as a percentage of the total area, and can usually range from 5 percent to 25 percent. (Low quarter DU looks at the driest 25 percent of the total area.) The following is an example of how the DU is calculated from actual catch can tests:

Twelve catch cans are placed within the coverage area of one zone that has no overlapping zones. The zone is operated for 10 minutes and the catchment amounts are measured in inches and multiplied times 6 (10 min. X 6 = 60 min.) to determine the actual application rate in inches per hour. The application rate for each catch can is:

#1 - 0.50	#2 - 0.39	#3 - 0.49	#4 - 0.45	#5 - 0.47	#6 - 0.47
#7 - 0.42	#8 - 0.49	#9 - 0.43	#10 - 0.41	#11 - 0.48	#12 - 0.40

DU = 100 (LQ/AVG)
  (0.40/0.45 = 0.89) X 100 = 89 percent DU
DU = Distribution Uniformity
LQ = The average of the low 25 percent is the total (#2-0.39, #10 - 0.41, #12 - 0.40) 1.20 divided by 3 which equals 0.40 inches per hour.
AVG = Average of the total catchment amounts or 5.4 divided by 12 = 0.45 inches per hour.

The Scheduling Coefficient (SC) is another method of uniformity which relates the lowest application rate in an area to the average application rate without regard to the size of the area. SC was developed by the Center for Irrigation Technology, (California State University, Fresno, CA 93740-0018) along with irrigation industry representatives.

The SC uses a sliding window of a designated size (e.g., 2, 5 or 10 percent of the coverage area) that identifies the area with the lowest application rate. The total area average application rate is divided by the lowest window application rate. The ratio is always one or more, and is used to increase the controller run time to provide adequate water for the driest window.

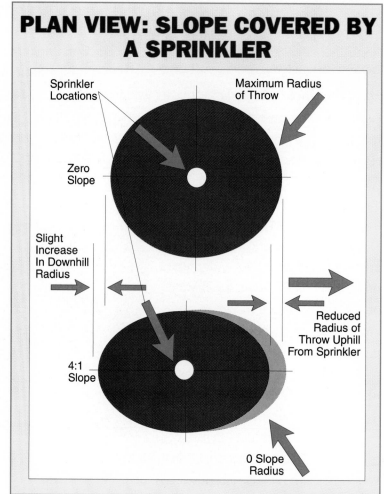

PLAN VIEW: SLOPE COVERED BY A SPRINKLER

Sprinkler Locations

Maximum Radius of Throw

Zero Slope

Slight Increase In Downhill Radius

Reduced Radius of Throw Uphill From Sprinkler

4:1 Slope

0 Slope Radius

The slope of a property can affect the uniformity of your sprinkler system.

The SC for the previous example of DU is:

$$SC = (AR/LAR)$$

$$0.45/0.39 = 1.15 \text{ Scheduling Coefficient}$$

SC = Scheduling Coefficient
AR = Average application rate of the area is 0.45
LAR = Lowest application rate is 0.39

Perfect uniformity produces a SC of 1. Any SC under 1.5 is considered good., but SCs of more than 2 are unacceptable. A SC value of 2 indicates the driest area is two times as dry as the average of the entire area.

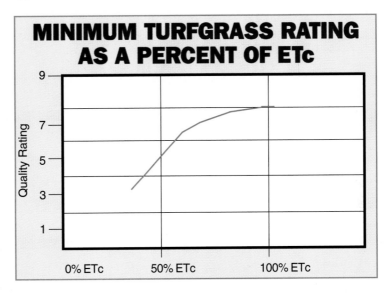

For a reliable test, use at least two catchment devices per sprinkler head, placing them two to three feet from each head and half way between the heads. Measurement of water application for each zone in the system using the "can" test method will provide for greater accuracy in water/time applications and will also conserve additional water. Measurement also increases your awareness of the uniformity of coverage provided by the irrigation system.

The Center for Irrigation Technology in conjunction with the Cooperative Extension, University of California, Riverside, has developed a new approach to optimize the SC. Visual quality observations of turfgrasses are made with ratings based on color, density, uniformity and vitality. Turfgrass quality is rated on a scale from 1 to 9, with 1 to 3 indicating superior quality. These quality ratings are indexed to the crop ET (ETc) rate to indicate the quality rating at various percentage levels of the ETc which will help turf managers more accurately predict irrigation run times that will produce the Minimum Turfgrass Rating (MTR) acceptable for the individual site.

Additionally, the operator should determine the Turfgrass Management Area (TMA) or the size of the dry spot that is acceptable. MTR, TMA and sprinkler pattern data allows the operator to more accurately predict run times that will produce the turfgrass quality defined by the operator.

The formula is:

$$\text{Set time} = \frac{(\text{ETc}) \times (\% \text{ETc}) \times (\text{SC}) \times 60 \text{ (minutes)}}{\text{Mean Application Rate (MAR)}}$$

Where:　　ETc =　　Daily crop evapotranspiration rate
　　　　　 %ETc =　　Minimum Turfgrass Rating as a percentage of ETc
　　　　　 SC =　　Mean application rate divided by the application rate within the
　　　　　　　　　 window (TMA) for the area with the lowest application rate.

Using the following conditions:

　　　　ETC　 = 0.30
　　　　%ETc　= 50% (MTR of 5)
　　　　SC　　= 1.64 (TMA is 10 by 10 and 4% of the coverage area)
　　　　MAR　 = 0.46

The formula appears as:

$$\frac{0.30 \times 0.5 \times 1.64 \times (minutes)}{0.46} = 32\text{-minute set time}$$

The operator has a run time of 32 minutes for the day, for a savings of 18 percent of the water required using the original SC method (39 minutes). This savings is possible while assuring that no dry spots exist of any consequence below a MTR of 5. This method combines uniformity data with turfgrass quality requirements in a more rational and realistic manner.

WATERING EFFICIENCY. Water application efficiency calculates the percentage of applied water that ends up in the root zone of the plant. Efficiency is the amount of water stored in the root zone divided by the amount of water applied. Properly maintained and managed irrigation systems can produce irrigation efficiencies above 80 percent.

The theories of uniformity and efficiency are completely separate. Systems can have a high uniformity and low efficiency, low uniformity and high efficiency or low uniformity and low efficiency. The goal is to strive for the highest uniformity (healthy appearance) and the highest efficiency (lower costs).

Several factors affecting system operating efficiencies are:

- Evaporation losses from spray droplets as they pass through the air.
- Wind drift that carries the droplet away from the irrigated area.
- Water overspraying on asphalt, concrete, structures and unnecessary areas.
- Water runoff from the turf surface region.
- Water that drains below the active root zone.

In 1942, J. E. Christiansen, a University of California professer and conservationist, published a method to explain and measure how uniformly water is distributed from a sprinkler. This number or percentage is referred to as the uniformity coefficient. Christiansen also studied direct evaporation loss from irrigation sprinklers in California and determined that more than 10 percent of the water may evaporate as it is applied in the afternoon during hot, dry periods, while the opposite occurred at night and losses were very low.

Additionally, a study of application efficiencies, indicates a wide range of efficiencies from

95 percent down to 65 percent. Water losses were either from deep percolation and/or evaporation. University of Arizona researchers K. R. Frost and H. C. Schwalen conducted tests to determine the effects of nozzle size, pressure, wind velocity, temperature and humidity. Frost and Schwalen concluded that water losses increase with higher temperatures, wind and operating pressure and decrease with increase in humidity and nozzle diameter.

Water losses are most directly related to the vapor-pressure deficit in the atmosphere which is dependent on the temperature and relative humidity. Near the saturation point or at a very low pressure deficit, water losses in a no wind condition were about 3 percent at 30 PSI pressure and noticeably greater at 50 PSI pressure.

Water application efficiency can range from 70 percent in hot, arid climates to 85 percent in cool, humid areas. It's important to remember that properly designed and maintained turf irrigation systems are capable of irrigation efficiencies at least as high as 80 percent.

Using a group of rotor heads operating simultaneously instead of a single head covering the same area will save up to 20 percent more water, according to a study by the American Society of Civil Engineers. One head cannot cool the air as quickly as five for the same sized area because the measured volume of water spray from five heads reduces temperature and increases humidity resulting in higher efficiencies.

IMPROVING EFFICIENCY. What other procedures can be performed to improve system efficiency? Anemometers (a spinning device with cups mounted on a pole that measures wind speed) can be installed with the irrigation system in order to measure wind and shut down system operation when wind velocities go above 10 or 15 mph. If prevailing winds are prevalent, it's recommended that the system be designed with head spacing for high winds or installed with rain shutoff devices on the irrigation system to avoid watering when it rains.

Other important steps include installing a temperature sensor that will turn off the system when temperatures drop below 38 degrees, particularly in cold climates. Additionally, soil moisture sensors are providing additional water savings that have not been possible in the past. These sensors read the moisture level in the soil and tell the automatic irrigation controller how much water to apply.

The Denver Broncos football team practice field has a soil moisture system that controls the amount of water applied. The turf appearance is great, and the usual problems with wet and dry areas are non-existent because turf manager Ross Kurcab selects times when the field is not in use and lets the sensors do the work.

If you're interested in more information on water uniformity, refer to *Irrigation, 5th edition,* available from The Irrigation Association, 8260 Willow Oaks Corp. Drive, Suite 120, Fairfax VA 22031. For more information on testing irrigation systems and water management, the one day Landscape Irrigation Auditor training program sponsored by the Irrigation Association (703/573-3551) is good.

In the last three chapters we have examined water management and the factors that affect good management practices. Soils, scheduling, pressure control, uniformity and efficiency must be understood in order to properly design, install, maintain and operate the system.

A nearly perfect irrigation system should have:

1. Uniformity of more than 90 percent.
2. Application rates that match the intake rate of the turf and soil.
3. Eighty-five percent efficiency.

QUESTIONS:

1. Does sprinkler head spacing affect irrigation system uniformity?

_____

2. What is an acceptable efficiency level for a pop-up spray head system?

_____

3. Can soil moisture sensors improve irrigation system uniformity?

_____

4. Can system uniformity be tested and measured in the field?

_____

5. Will poor uniformity of coverage affect the irrigation operating efficiency?

_____

6. What is more efficient, one head operating by itself or five heads operating as a group on one zone?

_____

7. What five things must be understood in order to properly design, install, maintain and operate the system?

_____

8. What is an Anemometer?

_____

9. Name two factors that affect system operating efficiencies.

_____

10. What is irrigation efficiency?

_____

(Answers to these questions are found on page 252.)

IRRIGATION DESIGNS...AND WATER

THE NEXT SIX chapters will focus on irrigation system design. Topics will include water sources, how to select the right equipment for a specific project, sprinkler head layout, zoning for control, hydraulics, pipe sizing, designing drip irrigation and a lot more.

Design is critical to the long-term success of any irrigation system. Design criteria result from repeated evaluations of existing systems, observations of installations and the feedback received from installation and maintenance contractors. It's also wise to monitor some design projects for water use and maintenance costs.

All of this information can be used to determine better methods for designing systems with improved performance. Discussion in this chapter will focus on the basics of design with an emphasis on what works, why it works, liability as a designer and contractor, xerigation — which is another term for water conservation — and efficient irrigation.

One of the most important elements of any properly functioning irrigation system is the quality of the water. Sand and other sediments can be very abrasive and may result in enlarged nozzle sizes, increased equipment tolerances and plugged nozzles and valve orifices. Chemicals in the water can cause deposits and scaling in equipment. Manganese, calcium deposits and algae can build up and plug these orifices/nozzles. Additionally, weed seeds, leaves and sticks can create maintenance headaches. Iron, as well as other minerals in the water, can stain sidewalks, structures and fences.

For example, shallow wells in Florida produce water that contains iron and other chemicals that corrode brass components, stain the concrete surfaces and structures and dramatically reduces the life expectancy of an irrigation system.

Water-efficient irrigation systems with low maintenance factors require clean water. Most landscape irrigation water comes from potable water supplies, which may not be as clean as you think. Flaking rust particles inside steel and cast-iron pipe which is used to deliver the potable water to the irrigation system can plug orifices. That is reason enough to avoid using

This "Y" strainer senses a reduced flow rate by measuring the pressure drop and opening a flush valve when too much sediment accumulates.

galvanized or steel pipe in an irrigation system, especially in areas where the system is drained in the winter allowing air to enter the pipe and potentially accelerating the corrosive process. PVC and polyethylene pipe will not corrode.

Other irrigation sources such as wells, lakes, rivers and ditches supply the balance of landscape water. Irrigation wells close to the seashore should be tested annually for salt water intrusion into the aquifer. When this occurs, the methods for filtration discussed here will not remove the salt from the water. Only a desalinization plant can do this.

Water pumped from streams or ditches often contains suspended solids and organic matter that can damage or plug the irrigation system. Likewise, water pumped from ponds and lakes may be high in organic content with algae, snails, bacterial slime, fish and clams. Take the Potomac River in Washington, D.C., for instance. Here snails can plug up orifices in heads and valves. All these elements can damage irrigation system components and reduce component life expectancy as well as potentially kill plants and stain structures.

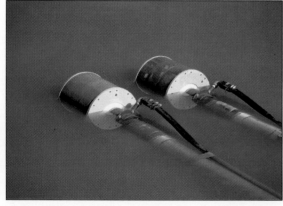

These two self-cleaning intake screens protect the pump and system from soil and debris.

How do you solve these problems and provide for adequate filtration? Whenever possible, it makes more sense to filter the water in almost any irrigation pumping system before it enters the pump. Filtering will extend the life of the pump impellers, bearings and bowls from two to eight years. If a filtering system is employed, the restriction of water flowing through the intake filter must be minimal, otherwise pump suction head will be too high causing a reduction in pump performance.

Pump efficiency drops and energy costs increase when there is friction loss or water pressure loss through the intake filter. Savings in electrical energy will be apparent over time and more than pay for the cost of an intake filter while preventing flow loss.

There are several self-cleaning pump intake filters available with little or no flow loss. When this type of filter is used, a small water line from the downstream side of the pump provides high pressure to blow debris off the outside of a drum-like screen that rotates. The unrestricted water intake of this filter does not affect pump efficiency and performance.

Centrifugal-type filtration systems are used widely for secondary filtration on the downstream side of pumps, or as a primary filter when potable water is the source. Centrifugal filters force the water into a downward spiral motion against the inside of the screen, moving the sand toward the bottom. Many centrifugal filters have self-cleaning flush valves that eliminate weekly or daily cleaning of the screen.

Sand media filters are excellent for the removal of organic material, fine sand and floating matter. Here water is forced through a tank partially filled with specially sized sand dependent on the type and size of particle to be filtered. The sand is cleaned by backwashing water through the tank. Sand media filters may not work well in areas where a large amount of particles must be removed from the water. Dirty water will result in frequent backwashing and reduce efficiency and performance.

Over-filtration can be costly. Specifying finer filtration than necessary causes greater pressure loss, reduced flow rate and more frequent cleaning. Greater pressure loss may result in a higher initial investment and higher pumping costs. A safe recommendation is to remove all particles larger than 1/6th the size of the smallest orifice in the system. If the orifice that transfers water to and from the area above the diaphragm in an electric control valve is 1/16 inch or 0.0625 inch, the ideal size of the filter screen is 0.0104 inch or 0.26416 mm (millimeters) — a 50-mesh screen — which is a U.S. designator for filter screen size.

Screen sizes are generally described as mesh sizes, but may be indicated in inches, millimeters or microns, etc. The following are some common mesh sizes and their equivalents in inches:

Mesh Screen Size	Inches
20	0.0330
100	0.0059
150	0.0044
200	0.0029

Before selecting a filter system, test the water to be filtered for organic material, suspended solids and sand in order to determine the type of filtration and the mesh screen size required. When using a sand media filter specify the size and type of sand.

Cheaper is not always better. Don't let price be the determining factor for filter selection. When you select a filter, look for maximum efficiency and make your selection based on the following criteria:

- Reliability
- Maintenance costs
- Particulate to be filtered
- Pressure loss through the filter
- Effect of water chemistry on filter equipment
- Pressure and volume required by the irrigation system

Designed flow rates for irrigation systems are sometimes much different than actual site conditions. Several years ago, when still in the irrigation system installation and service business, my foreman called me in a panic because a customer's sprinkler heads would not pop-up when the water was turned on. A brand new system that wouldn't work!

Needless to say, my stomach was full of butterflies while driving to her home. They remained there long after arriving on site and finding out she was right. Further adding to the mystery was the fact that six of us could pull up on a head and get the rest of the heads to pop-up and operate correctly.

The static pressure was about 70 PSI, so that wasn't the problem. We also checked all the valves in the system and the service line to see if they were closed. Limited restrictions in some locations were found.

After several days of investigation we finally dug up the service connection to the water main in the street and discovered that the 3/4-inch "K" copper line was kinked, restricting water flow. The mystery was solved, but there was no doubt that we lost money on that job. If you want to avoid a similar problem, perform a flow test at the point of connection or at the inlet to the backflow preventer prior to installation of the backflow device and the rest of the system. Time the flow into a 5-gallon bucket. If the bucket fills in one minute, the rate of flow is 5 gallons per minute (GPM).

Site water requirements must be determined prior to designing the system in order to size the system for adequate capacity. Important considerations are operating time constraints, turf traffic and the capability of the system to provide sufficient water for adequate plant growth during the summer when plant evapotranspiration (ET) is highest (peak demand).

Last year, we performed an irrigation system evaluation for a homeowners' association. It had 19 acres of irrigated bluegrass turf and the association could not keep it green during the heat of the summer; even with the system operating nearly non-stop. My first thought was that the water taps and meters were too small for the area. Five 1 1/2-inch water meters were interconnected with a 2-inch mainline. The static pressure was high at 85 PSI. Using the formula below, we calculated the peak weekly water demand for this site as follows:

$$Q \text{ in GPM} = \frac{\text{Area} \times 0.6234 \times \text{Weekly Water Requirement}}{\text{Days per week} \times \text{Hours per day} \times 60 \text{ minutes}}$$

Where:
Q = Flow in gallons per minute (GPM)
A = Area in square feet
0.6234 gallons = Volume of water in 1 square foot, 1 inch deep.

$$\text{WR (in inches)} = \text{Weekly turf water requirement divided by system application uniformity/efficiency}$$

$$D = \text{Number of days per week available for irrigation}$$

$$H = \text{Operating hours per day}$$

...using the following site information:

Area = 19 acres x 43,560 sq. ft. = 827,640 sq. ft.
WR = 1.5 inches/0.50 = 3 inches (Estimated system efficiency of 50 percent)
　　　　Difference between amount used and applied.
D = Six days per week
H = Eight hours per day

$$Q = \frac{827{,}640 \times 0.6234 \times 3}{6 \times 8 \times 60} = \frac{1{,}547{,}852.33}{2{,}880} = 537.45 \text{ GPM}$$

The above water requirement of 537 GPM is correct if all zone valves have the same flow rates. But we know that this rarely occurs in the average landscape irrigation system, so we added a flow variance factor of 30 percent (zone flow rates often vary because of changes in head quantities and flow rates) in order to meet system run time constraints. Add 30 percent to the 537 GPM and the maximum flow requirement equates to 698 GPM.

The next step in the process is to determine the capacity of the water meters. Using the American Water Works Association (AWWA) manual M-22, Sizing Lines and Meters, (page 45, column two) "Recommended Design Criteria — 80 Percent of Maximum Capacity" and derating the percent of maximum capacity to 70 percent (70 percent to 75 percent is the norm in the irrigation industry). The maximum flow rate for an 1 1/2-inch meter is 100 GPM; 70 percent represents 70 GPM.

Five 1 1/2-inch meters may provide 350 GPM, less the 30 percent flow variance, equals 269 GPM (269 + 30 percent = 349.7). The water requirement formula indicated a flow demand of 537 GPM plus a flow variance factor of 161 GPM (30 percent), for a total of 698 GPM and a weekly runtime of 48 hours. Changing the above formula the daily runtime for a flow of 269 GPM can be determined:

$$\frac{\text{Area} \times 0.6234 \times \text{WR}}{Q \times 60 \text{ minutes}} \ / \ 6 \text{ days} = \text{Hours per day}$$

$$\frac{827{,}640 \times 0.6234 \times 3}{269 \times 60} = \frac{1{,}547{,}852.33}{16{,}140} = 95.9 \text{ H} / 6 \text{ D} = 16 \text{ Hours}$$

Allowing a runtime of 16 hours a day, 6 days a week (one day off for mowing), the flow demand decreases from 698 to 269 GPM. The water supply was inadequate to allow watering within an 8-hour, 6-day watering window.

Next, we questioned if the water pressure was adequate. Static pressure was 85 PSI. We measured the operating pressure downstream of the backflow preventer at 60 PSI, or 25 PSI less than the static pressure (losses were higher then normal). Pressure requirements at the heads for this system were 45 PSI. That left15 PSI available pressure loss for the mainline, valves and lateral lines.

We suspected that the 2-inch PVC mainline was too small, resulting in high pressure losses. The average distance between water meters was more than 2,400 feet. The average flow per meter would be one-fifth of 269 GPM or 53.8 GPM.

Class 200 PVC 2-inch pipe with a flow rate of 54 GPM has a pressure loss of 1.66 PSI per 100 feet of pipe using the following Hazen and Williams formula:

$$0.2083 \ \times \frac{100 \wedge 1.852}{C \wedge 1.852} \ \times \ \frac{gpm \wedge 1.852}{D \wedge 4.8655} \ = \ \times 0.433 \ = \ P$$

Where: GPM = 54
C = Friction factor for smoothness is 150
D = Inside diameter in inches is 2.149
P = Friction loss in PSI

$$0.2083 \ \times \ (\frac{100 \wedge 1.852 \ = \ 5058.25}{150 \wedge 1.852 \ = \ 10,718.18} = 0.4719) \ \times \ (\frac{54 \wedge 1.852 \ = \ 1615.82}{2.149 \wedge 4.8655 \ = \ 41.35} \ = \ 39.08)$$

or 0.2083 x 0.4719 x 39.08 x 0.433 = 1.66 PSI

Half the distance between the meters was about 1,200 feet times the pressure loss of 2.26 per hundred feet equaling 27.12 PSI of pressure loss in the mainlines to the midpoint. Add losses of 7 PSI for valves and lateral lines and the losses are 34 PSI. Operating pressure was 26 PSI (60-34 = 26) in some of the worst case zones, which was 19 PSI below the recommended pressures of 45 PSI.

The two major problems with the system were the size of the water supply and the mainline. A larger water supply allows a more reasonable and shorter runtime. Larger mainlines increase the pressure at the head. As you can see from this example, restrictions in irrigation system capacity can result in significantly longer operating schedules (16 hours daily is a bit extreme).

Here's how you can check system capacity by reviewing the following factors:

The size of the area in sq. ft. (50,000)

$$Q \ = \ \frac{50,000 \ x \ 0.6234 \ x \ 3.0}{5 \ x \ 8 \ x \ 60} \ = \ \frac{93,510}{2,400} \ = \ 38.96 \ GPM$$

Days of the week available for watering   5
Operating hours available each day   8

The maximum weekly turf water requirement is typically divided by system application uniformity/efficiency. Generally, the maximum weekly turf water requirement will range from 1 to 2 inches. Check with your county extension agent or other expert in the area for your requirements. If you're not sure what the turf requirement is, start with 1.5 inches. Use an efficiency range of 30 percent at the low end for systems with poor efficiency up to 65 percent for pop-up spray head systems and up to 80 percent for rotor heads if the system was well designed and maintained. If you're not sure start with 50 percent. Using a weekly plant water demand of 1.5 inches and a 50 percent efficiency factor, the WR is 1.5 inches/0.50 = 3 inches of water applied per week.

The system capacity with a 30 percent flow variance factor is 50.65 GPM. Available operating time is one of the biggest factors affecting system operation, and many designers and contractors fail to consider system capacity during the design process.

Always check the operating pressure at the site prior to design, if this is not possible obtain the static pressure and calculate losses to the point of connection for the irrigation system. If site pressure is not available, contact the local water department for pressure at the site. If the pressure is too low, consider using sprinkler heads that have low operating pressure requirements and/or installing a booster pump if allowed by the water purveyor.

For the average system, if the operating pressure is more than 75 PSI, install a pressure reducing valve immediately downstream of the backflow preventer to control pressure and protect the system from damage.

CONSIDERATIONS FOR DETERMINING SYSTEM CAPACITY:

- Size of water source.
- Water pressure.
- Size of mainline and valves.
- Size of area irrigated.
- Pressure and flow losses.
- Window of operation: Watering constraints with athletic activities, mowing, pedestrian traffic, low water pressure and municipal water restrictions.
- System uniformity and efficiency, which is the difference between what the system delivers and the amount the plant uses.

QUESTIONS:

1. Which filtration screen size is most appropriate for nearly all types of drip irrigation: 20 mesh or 200 mesh?

_____

2. What is the required system capacity for a 6-acre site where the peak demand is 1.6 inches per week, the system efficiency is 60 percent and watering must be completed within 6 hours per day, three days per week?

_____

3. Does the municipal potable water supply ever require filtration?

_____

4. Will system flows increase over time when pumping large quantities of sand through the system?

_____

5. What size opening, in inches, does a 100-mesh screen have?

_____

6. Where do you find information on safe water flows through water meters?

_____

7. When pumping from lakes and canals, is filtration necessary, and if so, what type is necessary?

_____

8. Can filtration extend the life of the pump impellers and bowls?

_____

9. What is a safe way to determine filter size?

_____

10. Is a safety factor necessary when determining system capacity?

_____

(Answers to these questions are found on page 252.)

6

EQUIPMENT SELECTION

EQUIPMENT SELECTION, OFTEN done in haste, is critical to the long-term success of an irrigation system. Equipment that is high in quality and performance, low in maintenance, long lasting and water conserving are the essential components of a quality irrigation system.

Consequently, selecting the right equipment for an irrigation project is vital to the long-term success of the landscape and irrigation system. Particularly in view of what is often the case in this and other industries. Instead of considering the quality, performance and ease of maintenance of the equipment, many contractors and designers select equipment based on price or a personal relationship with a sales representative,

This chapter will identify key design issues and describe workable solutions to common design problems. Testing new products, when feasible, is recommended to gain firsthand knowledge of product effectiveness and efficiency.

SPRAY HEADS. Pop-up spray heads for turf areas should have a minimum pop-up height of 4 inches. This height is necessary because of higher mowing heights required for turfgrasses. A 6-inch pop-up height is advisable adjacent to streets where the turf builds up faster due to street sanding in colder climates.

The 12-inch pop-up height should be used in ground covers and flower beds. These heads require a heavy-duty retraction spring to prevent damage by maintenance crews. Some brands of heads have weak retraction springs or bad wiper seals, and will sometimes stick in the upright position.

A high quality wiper seal is necessary and important to prevent leakage around the stem during the irrigation cycle, to prevent debris from entering the head and to minimize blow by when the stem pops up. The blow by or flushing action that occurs when the water is turned on cleans the wiper seal and flushes debris from the head.

Heads should seal in the operating position at 10 PSI or less while the amount of water used

You can see the distribution profile of this head. The high uniformity of coverage is apparent in the even color and texture of the grass.

in the flush mode should be minimized to conserve water. The cap on the top of the head that allows access to internal parts should have a tight seal. Some heads have caps that leak and require tightening in the field.

Optimum operating pressure for most small sprayheads is 25 to 30 PSI. However, most of the sprayheads which have been installed are operating at pressures much higher, some as

This head needs a check valve to prevent low head drainage and erosion.

high as 110 PSI. Pressures above 40 PSI causes the spray to fog (the higher the pressure the smaller the droplet size) and may distort the spray patter. Additionally, the efficiency/uniformity of the spray is greatly reduced.

The head is a good place to control pressure in order to provide better uniformity within the zone/lateral area. A pressure regulator built into the bottom of the pop-up stem or the base of the head works well, and also prevents high flows in the event the nozzle is damaged or removed.

Pressure compensating devices that fit under the sprayhead nozzle are a maintenance nightmare. The major drawback with these devices is that they are designed for certain specific flow rates. (Different nozzles will require different pressure compensating devices.) This can be confusing for installers and maintenance professionals who may not get the right device installed with the nozzle. Also, these devices are not true pressure regulators; the downstream pressure will fluctuate whenever the upstream pressure changes. In the near future, most brands of pop-up spray heads will have pressure regulators installed in the base of the head. Rotor heads will, hopefully, follow.

Check valves are necessary in order to prevent water drainage out of the lateral lines several times every day when each watering cycle is finished. If the lateral line is partially filled or, even empty when the electric control valve opens (0.5 seconds), a surge or water hammer occurs in the line which can damage equipment and cause breaks in the system. Retaining the water in the lateral lines saves water and money, and lowers maintenance costs while extending the life of the system.

This impact rotor head was not intended for sites that are mowed with 36-inch width or larger mowers. The lids and bodies break and the body collects gravel.

ROTARY HEADS. On almost every existing irrigation project Keesen Water Management has evaluated over the last several years, a number of problems have been observed with impact pop-up rotor heads in plastic cases with light-duty plastic covers. The rotor heads are installed on projects where commercial mowing equipment is used. The result: broken covers and cases, cases filled with debris and impact heads that fail to retract after operating.

These heads have a surface area diameter of up to 8 inches which in turn exposes more area to vandalism and maintenance equipment. If the pressure is too high they operate like machine guns and the back splash ends up in the street or on nearby structures. Some people complain that they can't sleep at night because of the noise.

The solution to rotary head problems is the use of an internal drive, closed case rotor with wiper seals and spring retraction. Surface area diameters are reduced to less than 3 inches in diameter, and some of these heads can be installed below the surface of the turf in order to minimize damage from equipment and vandalism. Retractable rotors also minimize accidents and injury to passers-by. These heads generally have a higher pop-up height (up to 12 inches) and are also available with check valves and pressure control built into the head.

Rotor heads have a much higher uniformity of coverage than pop-up spray heads (70 percent to 80 percent vs. 50 percent to 60 percent). In the past, areas less than 30-feet wide required the use of pop-up spray heads. Small radius rotors (less than 30 feet in radius), however, are now used to replace the pop-up spray head in these small areas.

These two controllers are mounted in a weather/vandal resistant enclosure with lightning protection installed below the controllers.

Higher pop-up heights may be required in planting beds and along streets (sand build-up) in order to preserve good uniformity.

AUTOMATIC CONTROLLERS. Today's automatic controllers have numerous "bells and whistles" that can enhance system efficiency if understood and used properly. In reality, most systems are not operated properly because high-tech controllers can be intimidating, difficult to operate and hard to understand.

Several years ago we specified a new controller from a small company that produced only controllers. The controller was somewhat difficult to program and was repeatedly losing the intended program. The local distributor soon dropped support for the product and the controller manufacturer was sold to another irrigation products company.

This particular controller box was extremely small making it difficult to reach all of the #14 wires and the remote control pigtail. Even taking out the backup battery and replacing it was difficult. The problems experienced on the project caused both anger and embarrassment, and the owner requested that the clocks be changed. The moral of the story is twofold: buy equipment from a large reputable manufacturer with good local support, and attend to the clients' needs instead of making a buying decision based on a friendship with a manufacturers' representative.

The most important selection criteria for an automatic controller is ease of operation and simplicity. If an instruction booklet is needed to operate it, look for another controller. Most operators want to be able to walk up to the controller and quickly see how the program is set. Manufacturers have responded to this need and are producing new controllers that are easier to understand and program.

Lightning protection is the next most important feature to consider. If lightning is a problem in your area, it is advisable to have surge protection for the primary or 120-volt power source as well as on the secondary or 24-volt lines to the electric control valves.

Other important features that can contribute to irrigation efficiency are accurate timing, multiple repeat cycles (3 minimum), flexible day scheduling, sensor input and water budgeting. To apply the correct amount of water, timing should be accurate within seconds. Multiple repeat cycles reduce runoff and improve infiltration rates, while flexible day scheduling is useful for mowing, special events, etc., especially if water restrictions are enforced.

Water budgeting allows the user to set the rotor head zone timing for 1-inch of water per week or 100 percent coverage. At the beginning of the irrigation season the budget feature is set at 25 percent or 1/4-inch per week and then gradually changed as the season progresses. This feature saves time and encourages seasonal changes in water application. Water budgeting does not work well with pop-up spray heads, because the watering time per cycle should not exceed seven or eight minutes per cycle, and a one- to two-minute cycle may not be effective. If the controller has any pop-up spray head zones, my preference is to add or eliminate watering cycles instead of using the water budget feature.

Additional features that are useful for drip irrigation are multiple programs and timing in hours instead of minutes. If the program is stored in electronic memory, it is good to require memory retention of at least 10 years without power instead of replacing a rechargeable backup battery every year with pop-up spray heads.

PIPE & SWING JOINTS. Class 200 PVC is more than adequate for main lines if water pressure is properly controlled and Class 160 PVC works well for lateral lines in the areas of the country where frost will not enter the soil. 80# and 100# NSF polyethylene pipe, because of its flexibility, is well-suited to freezing climates. Avoid using non-NSF polyethylene pipe because the quality control may not be as good.

To protect the sprinkler heads from damage from mowing equipment it is important to use swing joints. Flexible tubing or premanufactured PVC swing joints are the best choice. PVC swing joints assembled on the job site have a tendency to leak because of the numerous threaded joints.

ELECTRIC CONTROL VALVES & BOXES. Look for valves that have an internal manual bleed so the valve box won't fill with water during manual operation at the valve. If the valve diaphragm tears, the valve should fail closed (this is a normally closed valve). Additionally, it is wise to attain a system with clog resistant, small valve ports and/or one that is self-cleaning. An encapsulated solenoid with a captured plunger helps the life span and reduces maintenance headaches.

Valves without flow control may be preferred because they contain fewer parts, resulting in fewer opportunities for mistakes. The flow control is often readjusted or sometimes completely shutoff by unknowing owners or vandals. Caution is the rule when using valves with a pressure reducing feature. My experience has been that few brands work well. Maintenance personnel are given little information on how to set these devices, and three years after installation no one knows what the pressure setting should be.

It's recommended that valve boxes be selected that are specifically designed to withstand potential compaction and weight against the sides and top of the valve box. Many valve boxes cave in on the sides (12-inch rectangular) because of a general lack in the structural strength of the product.

SUPPORT AND DISTRIBUTION. When selecting equipment it is extremely important to consider the local support and availability for any product. Maintenance personnel deserve and need local support for the products they maintain. A good irrigation distributor can provide in-house controller repair while the client has the use of a loaner or a replacement circuit board. Many distributors provide excellent help in problem solving and educational training programs.

Determine the requirements for your project and then make a selection from reputable manufacturers with local distribution. Products that work well can save money on maintenance, improve the reputation of our industry, conserve water and provide all of us with a beautiful, healthy environment.

QUESTIONS:

1. What are the essential components of a quality irrigation system?

_____

2. What type of head will soon replace the 15-foot pop-up spray head?

_____

3. What is the single most important criteria for selecting irrigation controllers?

_____

4. What is the optimum operating pressure for pop-up spray heads?

_____

5. What can be added to most irrigation heads to prevent low head drainage?

_____

6. What type of head provides the highest uniformity of coverage, pop-up sprays or rotors?

_____

7. What does the term "blow by" refer to?

_____

8. What two types of "swing joints" are the best to use under heads?

_____

9. How can the controller be protected from lightning?

_____

(Answers to these questions are found on page 252.)

SPRINKLER HEAD SELECTION AND PLACEMENT

TREES, SHRUBS, GROUND covers and turf all have different watering requirements. Consequently, the best type of equipment should be selected for specific applications.

Ask yourself this question: "Am I really doing the best design I can for my client, and can I improve what I'm doing"?

Most landscapers and irrigators can improve their design methods and should strive to do so on a continuing education basis. It will come as no surprise that almost every landscape irrigation project takes place in a complicated area. Because of the nature of the landscape — the slopes, narrow strips, obstructions and small areas — there are specific difficulties in every turf irrigation design.

Sprinkler head selection and proper placement is the single most important aspect of an efficient water conserving irrigation system. Those who design and/or install an irrigation system can be held liable if the system is not designed, installed and maintained (if appropriate) properly.

Before making a decision about head selection and placement, gather the necessary information to make intelligent decisions. Follow this checklist to help with decision making:

1. Potential water sources.
2. Water pressure.
3. Landscape planting plan or aerial photograph.
4. Topographical plans.
5. Site usage (athletic areas require more water).
6. Maintenance considerations and level of commitment.
7. Irrigation operating window.
8. Crop cover/soil characteristics.
9. Prevailing wind conditions and historical ET data.

Quality fittings between the sprinkler head and pipe could have prevented this head from blowing off.

HEAD SELECTION. First determine the available operating pressure range for sprinkler heads. Do not use an operating pressure which is within 10 PSI of the lowest available operating pressure because the water purveyor may decide to change the site pressure in the future, or population growth may increase demand while reducing pressure.

If a site has 75 PSI operating pressure at its low end and the high point on the site is 23 feet higher, 10 PSI will be lost in efforts to raise water to that point leaving 65 PSI. Subtract 20 PSI for system pressure losses and only 45 PSI is available at the base of the head.

The shape and size of the irrigated area should be analyzed next. If the site is a small residence with many confined areas, use a small radius rotor with an operating pressure of 15 to 30 PSI for areas that are wider than 15 feet. For areas that are less than 15 feet wide, small spray heads operating at the same pressure are adequate. If there are a lot of trees or other obstructions in the turf area, pop-up spray heads may be more effective.

If the site is a large park of 20 acres, with 65 PSI static pressure, 20 pounds of pressure loss, and a 5 PSI safety factor the design pressure equals 40 PSI. This limits the radius of coverage to about 45 feet. If the system static pressure is above 75 PSI with little elevation change, install a pressure regulating valve immediately downstream of the backflow preventer to protect the equipment.

TYPES OF HEAD SPACING

SQUARE	TRIANGULAR	RECTANGULAR

Diameters of coverage up to 120 feet are my choice for most projects. When the diameter increases beyond that, the water trajectory is higher resulting in more wind drift. The operating pressures must be above 60 PSI

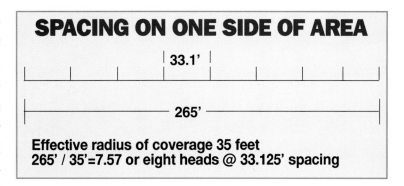

SPACING ON ONE SIDE OF AREA

| 33.1' |

265'

Effective radius of coverage 35 feet
265' / 35'=7.57 or eight heads @ 33.125' spacing

(requiring at least 80 PSI of static pressure) to spray water that distance.

In most cases, a pumping system or a booster pump is needed to obtain high pressures. Even with pumps it is better to keep the operating pressure lower in order to conserve energy over the long term. Avoid booster pumps because they increase maintenance and energy costs. Additionally, some water purveyors will not permit them on a site.

Low angle nozzles should be considered for the top of slopes, low overhead clearances such as mature trees or bridges and in high wind areas. If there are slope conditions, check valves in the base of the head are needed to prevent excessive low head drainage.

Trees, shrubs, ground covers, rose gardens and turf all have different watering requirements. The sprinkler application rate must be low enough to avoid runoff (although repeat cycles can be used to control this), yet high enough to complete irrigation within the required operating window. (See Chapter Two for more information on calculating precipitation rates).

Bubbler heads are great for rose gardens and other planting areas, but be cautious. Bubbler head areas must be relatively flat or erosion problems can occur because of the high application rate on a small area.

Recently, an evaluation was conducted of some landscaped medians in which the shrubs were irrigated by stream bubblers. The median was several hundred feet long with a 20-foot drop from one end to the other. When the contractor first operated the bubblers, soil erosion was a significant problem so he changed the stream bubblers (360 degree arc) to half circle nozzles to reduce the flow rate. This median is a major problem resulting in runoff, unhealthy plants, structural problems and water coming up through the concrete expansion joints in the street. A much better choice for this situation is drip irrigation, because of the low flow rates. (See Chapter Six for more information on head selection).

Spraying past the corners and eliminating the quarter head, results in dry spots, poor uniformity and wasted water.

SPRINKLER HEAD LAYOUT. The purpose of an irrigation system is to apply the water as uniformly as possible (see Chapter Four). Most manufacturers recommend a spacing of 50 percent of the effective diameter of the head. Specific heads may provide a higher uniformity at ranges from 40 percent to 60 percent depending on the individual head distribution profile, wind conditions and spacing configuration.

Spacing for any head should be no greater than the manufacturers' recommended spacing with some reduction of spacing for wind conditions.

There are three basic spacing configurations: triangular, square and rectangular. Rectangular spacing works best when prevailing winds (5 mph or higher) come from a constant direction. Triangular spacing is the most efficient for large area (more than one acre) irrigation, but for the average site, square and triangular spacing provide about the same uniformity. Square spacing is best suited for rectangular areas with 90-degree angles.

Head layout is another important consideration in excellent irrigation design. The irrigated area should be bordered, and a head or heads should be placed so they do not overthrow the area. Using a compass to draw the radius of spray, visualize the area of coverage for each head. The radius line depicts where heads should be spaced in relationship to one another (head-to-head spacing).

If there are 90-degree corners, start by placing quarter arc heads in corners, always avoiding half heads spraying over the corner. Measure the distance between the two quarters and choose a spacing that divides equally by the measured distance, and is equal to or less than the selected spacing. Or, if it's better suited, pick a different head radius. Use heads that have matched application rates between the various arcs and radii. Matched application rates occur when all sprinkler heads within a zone, regardless of the arc of coverage, have approximately the same application rates.

If the distance on one side of the irrigated area is 265 feet and the head radius is 35 feet, the head spacing will be 33.125 feet. If the perimeter is circular or curved, measure the length and calculate the equation. Place part circle heads around the entire perimeter and avoid

The heads at the top, middle and bottom of this slope are on the same zone, which creates a swamp at the bottom and dry areas on top.

This electrical transformer blocks 90 percent of the spray pattern from this quarter circle rotor.

overthrow beyond the perimeter.

After placing all the heads around the perimeter, full circle heads are placed inside the area from the edge to the center. If there are coverage problems in oddball areas, use part circle "back-up heads" where there may not be enough space for another row of full circle heads. Back-up heads are placed to water areas not receiving adequate overlapping coverage from other sprinkler heads.

Configure the irrigation system to avoid diffusing the spray pattern. This only reduces the radius of coverage by distorting the sprinkler pattern and lowering the uniformity.

OBSTRUCTIONS. Obstructions are anything which interrupt the spray patterns of an irrigation system such as trees, shrubs, fences, light fixtures and telephone pedestals. In situations such as these, smaller radius heads may be used to water around all sides of the obstruction. The more barriers such as mature trees, buildings, picnic tables, etc., and the shorter the area width, the smaller the radius required and will result in closer spacing and more heads.

SLOPES. Irrigating slopes presents its own difficulties: controlling pressure due to elevation changes, maintaining high uniformity and minimizing runoff. To some extent, slope problems can be resolved by setting heads on the slope at an angle that is halfway between vertical and perpendicular to the grade. For example, a 2:1 or 50

Flat sprays on low-angle nozzles are great for controlling the water, but be cautious of slopes and berms that can obstruct the spray pattern. Use 6- and 12-inch pop-up heads to properly place the water.

percent slope with a 26-degree angle results in a head tilted at a 13-degree angle. If the head is tilted any closer to vertical, erosion can occur. Setting heads at the appropriate angle allows more water to go further up the slope.

The uphill radius for a head with a 25 degree trajectory on a 2:1 slope is reduced by 30 percent and increases about the same on the downhill side resulting in an egg shaped pattern. The standard for head spacing on slopes is to reduce the spacing horizontally (heads at the same elevation) along the slope by 1 percent for every 1 percent of slope change beyond 20 percent grade or a 5 to 1 slope. A 50 percent slope, or 2:1, results in a 40 percent reduction in spacing across the slope. Consequently, a 30-foot radius head should be spaced at 18 feet across the slope with 26 to 30 feet between the rows depending whether the spacing is triangular or rectangular.

Vertical spacing on the slope is not changed (from top to bottom), except the spacing is increased between the bottom of the slope and the first row of heads because of the increase in the downhill spray pattern, and reduced as the last row interfaces with the top of the slope.

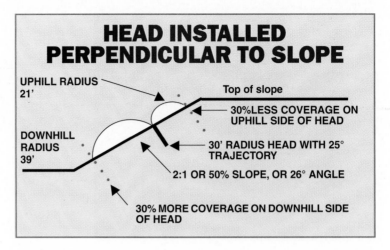

HEAD INSTALLED PERPENDICULAR TO SLOPE

UPHILL RADIUS 21'

Top of slope

30% LESS COVERAGE ON UPHILL SIDE OF HEAD

DOWNHILL RADIUS 39'

30' RADIUS HEAD WITH 25° TRAJECTORY

2:1 OR 50% SLOPE, OR 26° ANGLE

30% MORE COVERAGE ON DOWNHILL SIDE OF HEAD

When spacing heads on a flat plan surface remember to calculate the surface along the angle of the slope.

To do this, use the following formula:

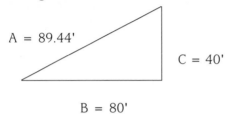

A = 89.44'

C = 40'

B = 80'

$A^2 = B^2 + C^2$
A = Distance along slope face
B = Distance of slope on plan sheet
C = Vertical rise in slope
For example, the plan dimension across a 2:1 slope measures 80 feet.
$A^2 = 80^2 + 40^2$
$6400 + 1600 = 8000$ (square root is $89.44 = A^2$)
The distance across the slope face is 89 feet.

The heads placed at the bottom of a hill or slope need to be zoned and valved separately. This ensures that less water will be applied to lower levels of the slope. Some water will always run off and irrigate the bottom of the hill. The midpoint of the slope will need a moderate amount of water; therefore these heads also require their own zone.

Finally, heads placed on the top of the hill will water the property for the longest period of time because this section will have no runoff from above, and is exposed to more sun and wind, thus increasing the evapotranspiration rate. In addition, heads with a lower angle of trajectory will increase the uphill and reduce the downhill radius improving sprinkler performance. Installation of lateral pipe

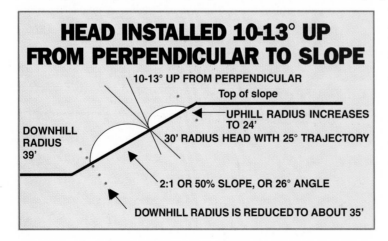

HEAD INSTALLED 10-13° UP FROM PERPENDICULAR TO SLOPE

10-13° UP FROM PERPENDICULAR

Top of slope

DOWNHILL RADIUS 39'

UPHILL RADIUS INCREASES TO 24'

30' RADIUS HEAD WITH 25° TRAJECTORY

2:1 OR 50% SLOPE, OR 26° ANGLE

DOWNHILL RADIUS IS REDUCED TO ABOUT 35'

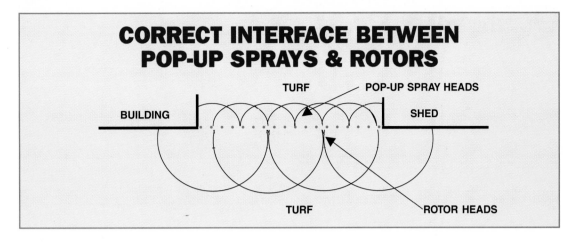

CORRECT INTERFACE BETWEEN POP-UP SPRAYS & ROTORS

should be placed horizontally on the slope to avoid major pressure variations and control erosion.

INTERFACING ROTORS WITH SPRAYS. Rotor heads and pop-up spray heads have different application rates so the interface must be bordered by both types to maintain good uniformity. Too often the tendency is to space the full-circle rotor head and the full-circle pop-up spray at the rotor head radius. Remember, the effective radius of coverage for a single head is 60 percent of its effective radius. Always draw an imaginary line between the two types, and border the line with both.

Public safety is another common problem irrigation designers should take into consideration. Heads should always retract to ground level. Likewise, spray heads in shrubs and ground covers should be 6-inches or higher and pop-up heads should be placed below sidewalk level and out of the way of pedestrian traffic and snow removal.

NARROW STRIPS AND MEDIANS. Define a narrow strip as an area which is less then 7 feet wide. If the area is less than 7 feet wide, strip heads can be used because of its specially designed nozzles which provide rectangular rather than circular coverage.

If an area is less than 6 feet wide, eliminate turf and ground covers in favor of tree and shrub plantings which can be watered by drip irrigation, or use sub-surface drip irrigation. Sub-surface drip irrigation usually consists of pipe with drip emitters installed at 12- to 24-inch intervals inside the pipe, or in the pipe wall. These pipes are installed 12- to 24-inches apart, and at 4- to 6-inch depths.

Small, irregular areas produce difficulties in head spacing. In some cases, you may have to use pop-up spray heads with a 5-foot to 8-foot radius nozzle.

In conclusion, many techniques enable an irrigation system to function properly and efficiently, even in difficult situations. When used appropriately, these techniques will also save water, provide for public safety and decrease liability. Using these techniques will save time, energy and costly mistakes.

QUESTIONS:

1. If the minimum available operating pressure at the head is 50 PSI, would a head with a 40 PSI operating pressure be appropriate?

_____

2. What are the three basic spacing configurations?

_____

3. When is it appropriate to use Bubbler heads?

_____

4. How much is a 15-foot head spacing reduced across a 3:1 slope?

_____

5. What is the single most important aspect of an efficient irrigation system?

_____

6. What is the surface distance (A) along a 2:1 slope when the plan distance is 100?

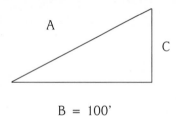

B = 100'

7. When should low angle nozzles be used?

_____

8. What is the first step in head selection?

_____

9. When should a pressure regulating valve be installed?

_____

10. What happens when the spray pattern is diffused to reduce the radius of coverage?

_____

(Answers to these questions are found on page 253.)

8

IRRIGATION'S LIFELINE: ZONING AND ROUTING

AN IRRIGATION SYSTEM will fail unless suitable zones and pipe routing are designed and installed properly. After completing sprinkler head layout, the next steps in the design process are to determine the size and area to be served by each zone in the system, the best location of the control valves and proper pipe routing. These steps are vital for controlling costs and improving system efficiency.

When developing zones, divide each site into areas of differing water requirements such as turf, planting beds, ground covers and so on. Cultivated planting beds require more water than areas mulched with humus or rock. Additionally, the height and density of the plant materials will provide shade for the soil surface, reducing the evaporation rate from the soil.

Many different watering requirements exist for every site. It's the job of the professional, to identify relevant terms to save water and improve the health of the landscape as well as to maintain its aesthetic value.

Be cautious when designing irrigation slopes or berms in which the slopes are greater than 4:1. When berms or mounds are present the top should be watered with a separate zone even if it consists of only one head. The peak of the berm will dry out much faster than the slopes, and will require additional water.

ZONING FOR CONTROL. Heads on the slope must be zoned separately from those at the peak or base of the slope. The heads at the base are zoned separately because of runoff and soil saturation from above. To meet the needs of the landscape, proper irrigation design requires that the operating pressure within a zone never vary more than 15 percent.

Site elevation changes will increase the pressure in the lateral line by 0.433 PSI for every

The turf and flowers have different water needs which require separate zoning. Also, the pop-up spray heads in the flower bed apply three times as much water as the rotors.

foot of elevation drop between the highest and lowest head. This may mean additional zones for the system unless pressure reducing devices (not flow controls) are installed at or inside the heads. If the elevation difference in a zone is 5 feet, the pressure at the lowest head is 2.17 PSI higher than the highest head. An 11-foot elevation change means an additional pressure of 4.76 PSI at the low head. If the heads are pop-up sprays operating at 30 PSI and flowing at 4 GPM, then some heads 11 feet below will be operating at a higher flow rate of 4.5 GPM.

This results in an increased flow of approximately 12 percent and a precipitation rate increase from 1.97 inches per hour to 2.17 inches, or 2/10 of an inch per hour of moisture. Conversely, an 11-foot elevation change means a pressure reduction of 4.76 PSI at the highest

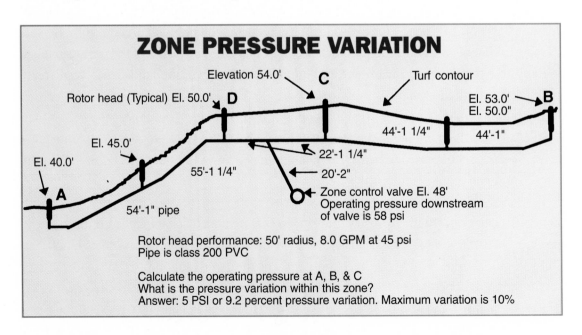

ZONE PRESSURE VARIATION

Elevation 54.0'   C   Turf contour

Rotor head (Typical) El. 50.0' D   El. 53.0' B
                                                         El. 50.0"

El. 45.0'

El. 40.0'   44'-1 1/4"   44'-1"

                   55'-1 1/4"   22'-1 1/4"

A   20'-2"

54'-1" pipe   Zone control valve El. 48'
                                     Operating pressure downstream
                                     of valve is 58 psi

Rotor head performance: 50' radius, 8.0 GPM at 45 psi
Pipe is class 200 PVC

Calculate the operating pressure at A, B, & C
What is the pressure variation within this zone?
Answer: 5 PSI or 9.2 percent pressure variation. Maximum variation is 10%

head. If the low head is operating at 30 PSI and a flow rate of 4 GPM, the head 11 feet higher will operate at a lower flow rate of 3.6 GPM minus pressure losses in pipe and fittings. When coupled with pressure losses in the lateral lines, a different zone is required when elevation changes exceed 5 feet or 6 feet in order to keep the lateral pressure variation within the required 15 percent.

This head should be moved out of the wetlands and up the slope to avoid spray diffusion.

Keesen Water Management recently designed a multi-family residential complex with 1:1 (45 degree angle) and 2:1 (26 degree angle) turf slopes. In some areas, the elevation difference from the toe (bottom) of the slope to the top was more than 15 feet, and the horizontal distance was 15 to 30 feet. Every 40 or 50 feet, a six-foot-wide stairway intersected the area for access to the street.

Walking, or climbing, up the slope was almost impossible yet maintenance crews managed to mow them somehow. It is particularly important that slope irrigation is designed to control the water application on the slope using separate zones when necessary. Instead of a design that is driven by costs, the physical limits of the zone (access stairs and walks) and ease of installation must be considered. This will also help turf maintenance crews by eliminating wet spots on the lower half and bottom of the slope.

Identify areas with varying exposures to sunlight. Areas on the north or east sides of a building or steep slope will require less water than the south and west side exposures, as well as any other areas of shade or sun. Also identify low places, drainage ways and storm water detention ponds, and zone these bottom areas separately from the adjacent areas.

Heads and valves should be installed adjacent to and outside of these low areas, not in the bottom where mud and water can affect equipment and ease of maintenance. Parking lot

This valve box is damaged frequently by vehicles and can be a liability if someone trips and falls.

medians and islands, due to the surrounding asphalt and heat, will require more water and a separate zone. Calculate all the head flows (GPM) for each identified area. Remember, all heads within the zone must have matched application rates, i.e., bubblers, rotors, spray heads and so on must be zoned separately. In addition, the nozzles within the zone should be matched.

Next determine the required system capacity in water flow (GPM) and pressure. (See Chapter Five on filtration and system capacity for information on how to do these calculations.) Once the maximum safe flow in GPM is determined, complete system zoning. If 30 GPM is the maximum allowable flow, then an area with a total flow of 101.5 GPM requires four zones at 25.38 GPM. Subsequently, three zones is 33.83 GPM which exceeds our maximum flow. Cheat on the maximum safe flow and it may come back to haunt you in high water costs, additional liability and water waste. When in doubt add another zone.

ZONE VALVE LOCATION. The most efficient location for the electric control valve (ECV) is in the middle of the zone, but because of elevation shifts, pressure controls and wire, pipe and trenching costs, this is not always cost effective for the entire irrigation system.

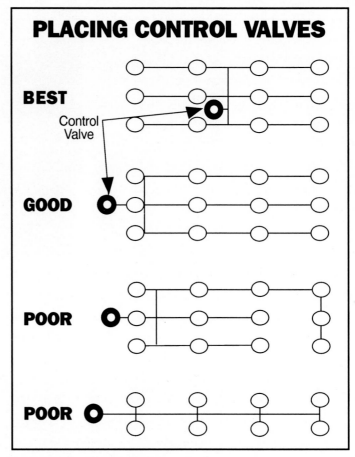

PLACING CONTROL VALVES

Cost effective control valve placement allows the valve to be on one side of the zone it serves. If an area is two zones in width, then it would be appropriate to route the main between the two zones. This will save on the cost of pipe and installation, as well as maintain a good balance of pressure throughout the zone.

The same is true at the end of the mainline where it is usually cost effective to stop the mainline prior to entering the last zone or two, depending on the distance and elevation change from the control valve and the closest boundary of the zone.

The mainline route, where the area widens and is more than two zones wide on any side, should have the mainline extended toward that area in order to better control the lateral pressure variation and lateral line surge. (Long straight lines have a much greater potential for surge then do shorter ones, and empty lines from low head drainage will increase the surge damage potential.)

End feeding long, single row zones with 30 to 40 pop-up spray heads can delay by minutes the time it takes for the first and last head on the line to pop-up. This affects the water distribution by placing more water closer to the zone. Center feeding the line will reduce surge potential and reduce the time between the first head pop-up and the last head pop-up.

Five years ago, we evaluated a parkway irrigation project in which most of the zones were composed of a long single row, several hundred feet long, with 20 feet of elevation change and the control valve located at one end. The maintenance personnel complained of both wet and dry spots throughout the turf area. The high pressure variation plus the elevation increase caused the wet and dry areas and poor uniformity of coverage. This was the first time we saw this theory in vivid reality.

If there are two or more rows of heads and little slope, shorten the rows and place several rows on the same zone. (Only if all heads have matched application rates.) Heads that are grouped together will cool the air more resulting in less evaporation and better compensation for wind direction changes during the watering cycle.

Valves and valve boxes should be kept away from walks, streets and driveways to avoid damage from vehicles and snow plows, lessen pedestrian liability, reduce visibility and prevent vandalism. The irrigation system should be as invisible as possible so it doesn't detract from the landscape.

MAINLINE & LATERAL PIPE ROUTING. Determination of zones and approximate control valve locations help route the mainline from the point of connection through the backflow preventer (if required) to the zone control valves and, in turn, the lateral piping to the heads.

The lateral piping is generally laid out in parallel rows, with up to six rows which are connected at the approximate center of the row with a header and then routed to the electric control valve (ECV). The pipe should be routed away from the control valve unless obstructions are in the way. Why waste pipe and pressure by directing the pipe back toward the valve? If you know that the pipe will be pulled into the ground then the route can be curved to the extent that the pipe and puller will allow. Routing will also be determined by the ease of construction and how the pipe and fittings are made.

Some designers try to place as much pipe as possible in the same trench and then branch off to the head locations. Any way its done, the amount of trenching is about the same, but single head branching will always increase the costs because of the additional fittings and repositioning of the trencher or puller.

In cases where trenchers can't get close enough to a building, branching is the only choice unless you want to dig it by hand. Trenching or pulling the pipe requires equipment that is several feet wide. Space must be allowed for this when routing the main and laterals.

Often, the easiest way to route the pipe is also an effective way to control pressure loss. The type of soil or rock will also have an effect on how the pipe may be routed. Minimize pipe routing through gardens and cultivated planting beds to avoid damage to lines. When routing the pipe, avoid obstructions and stay away from trees and their root systems.

Roots can cause polyethylene pipe to be squeezed shut, restricting flows and causing leaks. PVC pipe has greater rigidity and can better withstand damage from roots.

QUESTIONS:

1. At what slope ratio should the irrigation designer be concerned about control?

_____

2. What is the allowable pressure variation within a lateral zone?

_____

3. One foot of elevation change equals how much PSI?

_____

4. If the pressure within a zone is increased, will the precipitation rate increase too?

_____

5. What is the best location for a zone control valve?

_____

6. Is it a good practice to route the lateral line pipe through cultivated planting beds?

_____

7. How should a 4-foot high berm be watered?

_____

8. Should the calculated maximum safe flow rate be exceeded when zoning the system?

_____

9. Should single row zones with 30 to 40 pop-up spray heads be end feed?

_____

10. How is zone size determined?

_____

(Answers to these questions are found on page 252.)

ANALYZING IRRIGATION SYSTEM HYDRAULICS

IRRIGATION SYSTEM DESIGN and engineering have not always kept pace with equipment improvements. Consequently, pressure, flow calculations and hydraulic theories are often overlooked.

Have you ever designed or installed a system that had hydraulic problems? Was the pressure too high or too low in some areas? What about the problems you were never aware of like water hammer, a condition that reduces system life expectancy and causes leakage? If your designs have hydraulic problems, you could be liable for all repairs to the irrigation system and the property itself.

The key to a good design is to take the time to check the hydraulics for proper pressure and flow. The cost of water and energy make it imperative that the system be designed for efficiency and conservation.

Subsequently, keep the following water data in mind:

1. Water takes the shape of its container while seeking its own level.
2. Liquids are practically incompressible.
3. Water weighs 62.37 pounds per cubic foot and 0.036 pounds per cubic inch.
4. One cubic foot of water = 7.48 gallons of water.
5. One foot of head = 0.433 PSI.
6. One pound of pressure = 2.31 feet of head.

PRESSURE. Static pressure is an indication of energy that is available within the system when no flow exists. Static pressure is created by atmospheric pressure exerted on the water surface and the weight of the water above the point of measurement. It can also be created by pumping water into the system. Static pressure is measured in terms of a column of water exerting pressure through its weight at the bottom of the column, and measured as pounds per square inch (PSI). The formula is simple:

Pressure = Weight X height of the water
or
0.433 psi = 0.0361 lb./cu.in. x 12 cu. in. (height)

Elevation changes are a major influence on pressure and it will increase or decrease for every foot of elevation change at the rate of 0.433 PSI. You can determine elevation changes on a site by simply attaching a pressure gauge to the end of a hose and reading the high and

Too much friction loss in the pipe results in low pressure and a non-rotating head.

low pressure points with the gauge on the ground.

Determine the elevation difference by subtracting the low pressure (57 PSI) from the high (65 PSI) and multiplying the answer (8 PSI) by 2.31 which is 18.48 feet of elevation change.

Operating pressure, also referred to as dynamic pressure and working pressure, is the water pressure at various points within the system during operation. Changes in elevation and friction loss (pressure loss) from water flowing against the surface of its container will cause pressure to vary throughout the system.

The rougher the pipe surface the higher the rate of pressure loss. As water flow changes direction in fittings and valves, extra turbulence will cause additional losses.

FLOW PRINCIPLES. The flow quantity is the velocity or speed of the water and the area cross-section within the pipe. It is measured in gallons per minute. Flow velocity is a result of available energy to propel the water through the system and the acceleration change due to gravity.

Increased velocity results in a proportional increase in friction loss because there is a direct relationship between the quantity of water flowing and the velocity of flow. This is expressed as:

Quantity [cubic feet per second (cfs)] = Area [square feet (sf)] x velocity [feet per second (fps)], which can be simplified to:

$$\frac{\text{Quantity (cfs)}}{\text{Area (sf)}} = \text{Velocity (fps)}$$

The actual formula would appear as:

$$\frac{\text{(Flow in gpm) x (1 cubic foot / 7.48 gal) x (1 min / 60 sec)}}{\text{((ID\^2) x PI) / 4 x (1 cubic foot / 144 cubic inches)}} = \text{Velocity (fps)}$$

Where: ID = Inside pipe diameter
          PI = 3.1416

For example, the velocity of flow through a 2-inch CL 200 PVC pipe (2.15 inch inside diameter) with a volume of 60 gpm is:

$$\frac{[60 \text{ gpm x (1 cu. ft. / 7.48 gal.) x (1 min. / 60 sec.)} = 0.1337]}{[((2.15\^2) * 3.1416) / (4 * (1\^2 / 144\^2)) = .0252]} = 5.31 \text{ fps}$$

Several years ago we did an evaluation for a medical facility that was situated on the side of a hill with 90 feet of elevation change. Drawings of the system were non-existent, four 1 1/2-inch meters served the system and all of the mainline was 2-inch PVC. The system was in a constant state of activity to keep up with the plant water requirements. At first, we thought the water supply was too small, but after checking the site peak demand requirements we determined that the four meters were adequate.

As the evaluation continued, we checked the equipment and measured the operating pressure in each zone. We then compared the operating pressure of the various zones (8 PSI to 86 PSI) and made allowances for differences in elevation and distance from the sources. We discovered that the operating pressure in this interloped 2-inch mainline was 20 PSI lower because of friction loss at the furthest point from the source without any adjustment for ground elevation.

The contractor probably used 2-inch mainline on all of his commercial projects regardless of friction loss, velocity or the size of the project. As the water moved through the mainline, the pressure kept dropping resulting in inadequate pressure for a portion of the system.

The exception occurred where the mainline was much lower than the source and the pressure was higher, resulting in velocities that were damaging the system.

As water flows through the pipe, the flow is either laminar or turbulent depending on the velocity. Laminar flow occurs when the particles of the water follow separate non-intersecting paths with little or no eddying or turbulence. Turbulent flows see the water swirling and rotating as it moves through the pipe. Laminar flows are generally velocities of 1 to 2 fps and less.

FRICTION LOSSES. Whenever water is flowing in the system, there is a loss of pressure due to friction with the pipe. Friction loss results

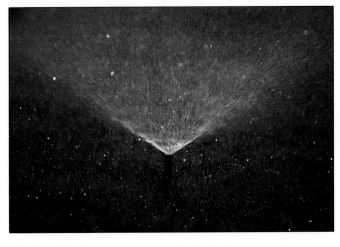

With an operating pressure of 30 PSI, this head is performing at its best due to correct friction loss calculations.

in an accumulated loss of pressure as the water moves through the system. As the velocity increases the friction loss also increases. Friction loss may be reduced by increasing the size of the pipe, reducing the flow rate, reducing the velocity or using a smoother material such as PVC vs. steel pipe.

Most of the pipe friction loss charts are based on the Hazen-Williams formula. Each type of pipe is categorized by the type of material and given a "C" value which indicates its relative smoothness or roughness. The "C" value for new PVC pipe is 150, for new copper and poly-ethylene 140 and for steel pipe 100. (See Appendix.)

The PVC pipe surface is much smoother than the steel pipe. As the pipe ages, rust can occur in steel pipe, corrosion may happen in copper pipe and suspended solids and sand can cause abrasion in any pipe which will reduce the "C" value and increase the friction loss as the system ages.

The Hazen-Williams formula reads:

$$0.2083 \times \frac{100^{1.852}}{C^{1.852}} \times \frac{gpm^{1.852}}{D^{4.8655}} = \times 0.433 = P$$

Where:
- C = Friction factor for smoothness
- D = Inside diameter in inches
- P = Friction loss in PSI per 100 feet

Friction loss for 100 feet of 2-inch CL 200 PVC pipe with a flow of 60 gpm is:

$0.2083 \times [(100^{1.852} = 5058.25) / (150^{1.852} = 10718.18) = .472)] \times [(60^{1.852} = 1963.98) / (2.15^{4.8655} = 41.46) = 47.379)] \times 0.433 = 2.017$ PSI of loss.

Many people use the friction loss charts provided in design manuals, but Keesen Water Management prefers simple computer programs written to solve these flow loss problems when determining velocity. It is easier to make errors using charts provided in manuals by reading the wrong column, size, chart or picking the velocity instead of friction loss.

EXTREME PRESSURE. Surge pressure or water hammer can damage irrigation systems and reduce system life expectancy. Surge pressure is a series of pressure pulsations of varying magnitude, above and below the normal pressure in the pipe. The magnitude and frequency is dependent on the velocity of flow, size, length and material of the pipe. Shock results from these pulsations when the flow of water is stopped in a short period of time.

Surge is sometimes accompanied by a sound comparable to a hammer struck against a pipe in copper and steel pipe. Intensity of sound is no measure of pressure magnitude. Tests show that if 15 percent of the shock pressure is removed by surge absorbers installed in the line, the noise is eliminated but relief from the surge is not enough to protect the system from damage.

Valve closure time is the other key factor affecting surge pressure. Most irrigation system electric control valves have a closure time of less than one second. The Center for Irrigation

Technology at CSU Fresno, Calif., has tested numerous valves and found the actual closure time to be 0.5 to 0.8 seconds.

Surge pressure can be calculated:

$$S = \frac{V \times L \times 0.07}{T}$$

Where:

S = Surge pressure (additional pressure over normal operating pressure
V = Velocity
L = Length of straight mainline
T = Time of valve closure in seconds

Examples:

10 fps x 400 feet x 0.07/0.5 seconds = 560 PSI surge pressure
5 fps x 400 x 0.07/0.5 seconds = 280 PSI
5 fps x 100 x 0.07/0.5 seconds = 70 PSI

This is surge pressure only, actual pressure on the system would be surge pressure plus operating pressure.

The surge pressure increases proportionally with additional length and velocity. Directional changes within the system will help dissipate this energy, but if the surge is too high, elbows may crack or fittings may be blown off of the pipe. Installing control valves on a riser and elbow above the mainline will help suppress some of the surge.

It's best to always maintain velocities under 5 fps in mainline designs. Lateral lines that are drained after each operation also experience surge, and the velocities should be kept under 6 fps.

Sprinkling in the wind: These missing nozzles could be the effect of high pressure and surge in the lateral line.

QUESTIONS:

1. Will increased velocity result in a proportional increase in friction loss if the pipe size remains the same?

2. Is dynamic pressure an indication of energy that is available within the system when no flow exists?

_____

3. What is the static pressure at the base of a water tower with the water level at 145 feet above the base?

_____

4. Are laminar flows common in most irrigation systems?

_____

5. Does copper pipe have a smoother inside surface than PVC pipe?

_____

6. What is surge pressure?

_____

7. What does the "C" value indicate?

_____

8. What is the friction loss for 150 feet of 2-inch CL 200 PVC pipe with a flow rate of 55 gpm?

_____

9. What is the velocity of water flowing through a 1-inch CL-160 PVC pipe with a flow rate of 15 gpm?

_____

10. Is valve closure time usually less than one second?

_____

(Answers to these questions are found on page 253.)

SAFELY SIZING THE IRRIGATION SYSTEM

POORLY SIZED IRRIGATION systems can result in problems for years to come. But these difficulties can be avoided with attention to water flow and related external factors. Sizing the system pipe and valves is critical to its function, and can be a liability to the designer if the system fails to perform properly. Improper pipe and valve sizing can result in low pressure and inadequate coverage, wasted money invested in oversized equipment and water hammer which reduces the life expectancy of the equipment.

Many designers are unaware of the problems they have created with their irrigation designs. If the contractor discovers the problem before installation, he will often try and correct the mistake in the field. In other cases, he will attempt to correct the problem after the system is installed. Designers should ask contractors for feedback on individual designs and explore the problems which were encountered during the installation of the system to avoid future problems.

You should use the hydraulic calculations worksheet provided in the Appendix, before beginning the sizing process. The worksheet includes:

- Valve number and flow rate
- Available water pressure
- Loss or gain due to elevation
- Detailed list of various component pressure losses and velocities
- The head operating pressure requirements
- Project location and date

WATER SERVICE & METER. Determining safe flows for service lines and meters is critical to the success of any design. The smaller water services and meters (1-inch and smaller) on municipal water mains are significantly more restrictive than larger services and meters. Pressure losses are much higher in the 3/4- and 1-inch service lines and meters. These losses

Accurate irrigation results from good, original designs. Designers and contractors should work together to achieve desired results.

can severely limit the flow of water into the system. Use extreme caution if the service line is galvanized or lead pipe. This type of pipe is old and usually corroded, resulting in very high pressure losses, very low flow rates and systems that won't work.

About 25 years ago, we installed a small residential irrigation system with sprinkler heads that, unfortunately, would not pop up when the water was turned on. With several people each pulling up a head we could get the rest to seat and spray, but the coverage was unacceptable. We checked the valves to be sure they were open then called the water department for information on the type and age of the service. We were dealing with an old 1/2-inch lead service line from the main to the meter. The only choice was to replace the old service line with a new 3/4-inch copper line.

We should have checked the supply before the system was installed and warned the owner, or designed for a lower flow rate. As it turned out, the owner indicated a problem with water pressure when he took a shower. After much convincing, the owner decided to pay half the cost of the new service and we paid the rest.

The best guide for sizing water services and meters is the "Manual of Water Supply practice and Sizing Water Service Lines and Meters, AWWA M22" from the American Water Works Association (303/794-7711). Water flow through meters is limited by the maximum flow capacity of the meter and maximum pressure losses recommended in the manual. The selection and size of the water meter should be based only on flow requirement, not on pressure loss through the meter.

The maximum design capacity, as recommended in the AWWA manual, is 80 percent of the maximum safe flow when the water meter serves only the irrigation system. The limiting factor is the affordable pressure loss in the service line.

Although many designers use flow rates up to 75 percent of the maximum safe flow, we prefer to use 70 percent for systems with existing water meters. For example, a 1-inch water meter at 70 percent of 50 gpm is 35 gpm.

When the water meter jointly serves the irrigation system as well as indoor water consumption and other uses, remember to consider the customers peak demand. Don't use all the available water for irrigation. Size the meter at 50 percent or less of the AWWA maximum safe capacity for irrigation demand, and schedule irrigation operation during the night hours. (See Chart "Recommended Maximum Meter Sizing For Joint Use".) This sizing method reduces pressure losses through the meter and will prolong the meter's life expectancy. Because of the high flow losses in the 3/4- and 1-inch service lines, it is impractical to design this size system at 80 percent of the maximum flow capacity.

Some water purveyors require the service line and meter to be the same size. The 1-inch and smaller service lines are considered the restrictive factors in sizing, not the water meter. According to the AWWA manual, the service flow should not exceed 15 fps velocity and extreme caution should be used when service velocities are this high. Some water purveyors have adapted this standard while others use 7.5 fps depending on the pressure rating of the service pipe. Urban water distribution systems have pressures ranging from 35 to more than 150 PSI with averages in the 50 to 80 PSI range. Excessive pressure loss in the service line and meter cannot be tolerated with pressures in this range.

For example: A system with a 1-inch service (50 feet) and meter, a 30 PSI operating pressure requirement at the head and 60 PSI of static pressure, the total allowable loss could be as high as 30 PSI. Keep the allowable loss below 20 PSI to allow for pressure fluctuations and possible future modifications. When using a 70 percent of maximum flow criteria a flow of 35 gpm is possible through the meter, but the pressure loss in 50 feet of 1-inch K copper tubing is 20.6 PSI with a velocity of 14.4 fps. This means that the velocity and pressure losses are too high and should be cut back to about 18 gpm for a PSI loss of 6 PSI and a velocity of 7.4 fps.

Low pressure at this head is the consequence of too much friction loss in the mainline.

RECOMMENDED MAX. METER SIZING FOR JOINT USE

Size in.	Flow for Irrigation 50% of Maximum Capacity		Type K Copper (C = 130) Service Loss & Velocity		
	gpm	psi loss	psi/100'	psi/50'	vel/fps
5/8 X 3/4	10	3.7	16.4	8.2	7.4
3/4	15	3.6	34.8	17.4	11.0
1	25	3.7	21.9	11.0	10.3
1 1/2	50	4.9	11.4	5.7	9.3
2	80	4.9	7.0	3.5	8.5

This is safe for most situations and allows for 14 PSI of loss in the other system components as well as a safely sized design.

BACKFLOW PREVENTER & SPECIALTY VALVES. Industry standards for water flow through backflow preventers is a maximum of 7.5 fps. A 2-inch reduced pressure backflow device has a maximum flow rate of about 77 gpm at 7.5 fps. A 1-inch pressure vacuum breaker at 7.5 fps is about 20 gpm. Many designers ignore this requirement or are unaware it exists. Check the manufacturers' pressure loss charts. Never exceed the manufacturers' recommended maximum velocity for backflow prevention.

When sizing the electric control valve, generally reduce the dimension one pipe size below the outlet pipe. This reduces the cost of the valve with a relatively small increase in pressure loss of 2 to 4 PSI. If the static pressure is more than 70 PSI, the loss is easily affordable. When using pressure reducing valves sizing should be based on the valve's flow capacity at a reduced pressure level. Check the manufacturers' catalog for detailed flow information. For most commercial and residential projects, install pressure reducing valves whenever the static site pressure is above 75 PSI. Be sure and check the flow rate to avoid oversizing valves.

MAINLINE & LOOPED PIPE SIZING. When sizing the mainline it's crucial to consider the potential surge pressure that can be produced from fast-closing valves. This surge danger can be reduced by maintaining velocities under 5 fps. Keesen Water Management designs mainlines for velocities that are always less than 5 fps, with the average project at 3 to 4 fps. Staying in this 3 to 4 fps range greatly reduces the chance of surge problems, and enhances the life expectancy of the system.

Working from the furthermost valve back to the source, size the pipe to maintain velocities that are about 3 or 4 fps. Don't forget to adjust for possible limitations resulting from affordable mainline pressure loss. If the static pressure is 52 PSI and the system requires 30 PSI at the head, the affordable loss is 22 PSI less a safety factor of 5 PSI; a total of 17 PSI. This might require larger sized piping in mainline and laterals in order to minimize the loss in the system.

Many systems have mainlines that are looped and/or interconnected to balance the pressure and flow. When feasible, these looped mainlines allow smaller pipe sizes or lower pressure losses because two pipes will usually have a capacity of 1.5 to 2 times the amount of water as one the next size up. As the water travels to the control valve it will follow the path of least resistance with most of the water flowing in the pipe segment where the valve is closest to the source while the balance will flow in the other segment of the loop. The pressure loss is usually calculated as 1/2 the flow through 1/2 of the total loop length.

The flows will adjust to create equal pressure loss in each leg of the loop. Looping the mainline will usually reduce construction costs and can provide for a better system. When the mainline losses are calculated, tally them and add for PVC fitting losses. Fitting losses are calculated using equivalent pipe lengths from the table "Pressure Loss in Pipe Fittings and Valves" in the Appendix. A 2-inch 90 degree elbow is equivalent to 7.4 feet of 2-inch pipe and can be calculated with the same flow rate as the adjacent 2-inch pipe. Additionally, the length can be added to the pipe footage and losses calculated together as long as the flow rates are the same.

SIZING LATERAL LINES. The industry standard for pressure variation in the lateral lines is 20 percent of the highest pressure in the lateral. This means that the pressure difference between any two heads in the zone should not exceed 20 percent of the highest pressure within the zone. This standard is necessary to provide high uniformity and efficiency. If the required operating pressure is 30 PSI, the allowable pressure variation is 6 to 7 PSI. If the required pressure is 50 PSI, the allowable variation is 10 to 12 PSI. To provide a higher uniformity of coverage and a more efficient system keep the pressure variation in the lateral line at less than 5 PSI.

Start at the head that is the greatest distance from the control valve and measure the distance to the next upstream head or tee in the line. If the flow is 4 gpm at the last head and the distance to the next head is 15 feet, calculate the pressure loss using the formulas or pressure loss charts in the previous chapter. The amount of loss in 3/4-inch polyethylene is 1.62 PSI per 100 feet or 0.24 PSI of loss.

Move to the next upstream head or tee and add the flow of that head to the previous one. The flows are accumulated as the losses are calculated. Next, proceed to the valve. The second head has a flow rate of 2 gpm, so 6 gpm is used in calculating the losses in the next upstream segment. When you come to a tee in the lateral line, determine the entire flow for that segment of the lateral and add it to the total flow amount for calculating the loss in the next leg back to the valve.

Write the flow, velocity, distance and PSI losses on the hydraulic calculations worksheet (see Appendix) for a permanent record. After all the legs are calculated, tally them and add for fitting losses to determine if the losses are acceptable. When figuring the system pressure losses, always check the valve furthest from the water source and the zone with the highest flow. Using the zone with the highest flow rate, finish calculating.

Another method of sizing the lateral is to limit the quantity of water flowing in each size of pipe. For many years the following flow rate limits for polyethylene laterals have been successfully used with less than 30-foot spacing between the heads:

Size Poly Pipe	Max Flow	PSI loss per ft	Velocity fps
3/4"	6 gpm	0.034	3.6
1"	14 gpm	0.051	5.2
1 1/4"	24 gpm	0.036	5.1
1 1/2"	33 gpm	0.031	5.2
2"	55 gpm	0.024	5.3

Lateral lines on slopes should be installed along the contour rather than up and down the slope. This will prevent pressure variations due to elevation changes. Acceptable pressure losses for most irrigation system (with adequate water pressure) components are shown in this chart:

component	PSI loss ranges
Service and meter	6 to 10
Backflow preventer and control valve	6 to 12
Mainline	2 to 5
Lateral lines	3 to 5
Totals	17 to 42

Finish filling out the hydraulic calculations worksheet with pressure and elevation information. Perform calculations for the valve furthest from the water source and the zone with the highest flow. Using the zone with the highest flow rate determine if the pressure losses are acceptable. If not, upsize the pipe or valves to allow a minimum safety factor of at least 5 PSI.

Proper system function is dependent on accurate hydraulic calculations for every system designed.

QUESTIONS:

1. Which size service lines and meters have higher pressure losses?

_____

2. What is the best guide for sizing water services and meters?

_____

3. What is the industry srtandard for water flow through backflow preventers?

_____

4. Why do many designers size the control valve one size smaller than the downstream piping?

_____

5. When should pressure reducing valves be installed for most commercial and residential projects?

_____

6. What is the advantage of a looped mainline?

_____

7. How should the meter be sized when the water meter jointly serves the irrigation system as well as indoor water use?

_____

8. How are fitting losses calculated?

_____

9. What is the industry standard for pressure variation within the lateral line?

_____

10. When determining total system pressure losses, what two zones should be calculated?

_____

(Answers to these questions are found on page 254.)

11

SUCCESSFUL DRIP IRRIGATION

IN APPROPRIATE LANDSCAPE settings drip can now be used with a high degree of confidence when designed, installed and operated properly. Drip irrigation is used to directly apply water into the soil or at the soil surface above the active root zone of plants. Its purpose is to efficiently irrigate plants in situations where water would be wasted if sprayed in a conventional manner.

Drip emitters disperse water at the emitter location or with the use of 1/8-inch capillary or "spaghetti" tubing. Water is distributed from a single or multi-port emitter to various points above or within the plant root zone at a flow range of 0.5 to 24 gallons per hour (gph). The emitters should be spaced to provide wetting of 70 percent of the active root zone.

Drip is synonymous with water conservation and Xeriscape.™ Drip, if operated properly, is one of the most efficient means of irrigating and can provide the landscape with many benefits. Advantages include:

 • Healthier plants, longer lasting blooms and less disease result from a slow application of water that prevents soil saturation while improving the oxygen level in the soil. Additionally, the foliage is not soaked by irrigation.

 • Reduced soil erosion in which the drip rate is less than the intake rate of most soils, and runoff is non-existent.

 • Lower liability from runoff and spray on roads and sidewalks, and fewer personal accidents caused by conventional irrigation equipment.

 • Low flow rates allow for the use of irrigation close to structures when properly operated, and reduces drip system costs.

 • Inexpensive temporary system that can be abandoned after two or three years.

 • Anytime operating hours that won't interfere with activities or traffic.

 • Less expensive than conventional systems.

 • Lower maintenance costs than conventional overhead irrigation, if properly designed.

• Pays for itself in water savings alone — usually within three to five years.

• Reduced vandalism specifically when lines and emission points are installed below grade in the soil.

Drip irrigation equipment and methods have improved dramatically over the last 10 years. The disadvantages of drip are rapidly disappearing and the level of maintenance is greatly reduced. The phrase

Water dripping from the bug cap at the end of the staked capillary tubing is one of the most efficient means of irrigating and provides many benefits.

"I can't see it operate" is a lazy excuse for not checking the soil moisture level or visually checking plants for potential stress and disease.

Keesen Water Management has been designing drip for more than 10 years, but not without initial problems. Fortunately, better techniques have been developed over the years enabling designers to draw and implement successful drip systems with little or no maintenance. Spray-type systems usually require more maintenance than drip systems.

Merle Moore, the horticulturist for the Denver Zoo, used drip emitters and dripper lines (porous pipe) to irrigate the indoor "jungle" in the new Tropical Discovery exhibit. Porous pipe is used to irrigate vertical "plant walls" and drip is used in plant pockets and inside exhibits.

Avoid the use of micro sprays installed on 1/8-inch plastic risers. The micro sprays blow off and easily bend over.

The use of this system provides viewers the illusion of lush growth without the threat of water waste and maintenance headaches.

DESIGN. Design is critical to the success of a drip irrigation system. Selection of quality drip equipment that meets the needs of the plant is the first step in the process. Additionally, specifying self-cleaning emitters and providing for excellent water filtration are the most important factors in the design process.

Five years ago, when evaluating a problem drip site at the request of a landscape contractor, we observed the system in operation and found it producing a steady stream of water at the emitter outlet instead of small drops of water. It was equipped with pressure compensating, self-flushing emitters. If half of the emitters were plugged or half of the emitters were removed from the pressure reducing device, the system dripped properly — otherwise it didn't. This problem occurred because the drip tubing line was undersized, and the contractor was not aware of the need to size the valves and drip piping to accommodate the flush mode.

Self-flushing emitters flush at the beginning of every drip cycle (a steady stream of water is visible instead of drips). This continues until the pressure builds up in the drip lateral line and the emitter goes into the drip mode. During the flush cycle, the emitter requires up to 750 percent higher flows to flush than the gallons per hour indicated in the catalog for basic drip operation.

FILTRATION. This is required of any water source to ensure the longevity and uniformity of the drip system. Historically, the industry has installed a strainer and a pressure regulating valve at the zone control valve. Some multi-outlet emitters have built-in, serviceable strainers to protect the emitter from plugging and to save the cost of installing a strainer at the control valve. For ease of maintenance, always install a filter at the control valve location. We prefer a minimum filtration size of 150 mesh. (See Chapter 5 on filtration.)

PRESSURE. We consulted on a project in Denver in which the pressure reducing valve was installed downstream of the control valve. The maintenance contractor complained that some of the drip components were coming apart. We investigated the problem and, with the help of the manufacturer, determined the cause.

Because the pressure reducing valve was installed downstream of the automatic control valve it took some time for the pressure reducing valve to set at the correct pressure. This resulted in a downstream surge which literally blew the emitters out of the drip tubing. When pressure reducing valves were installed upstream of the automatic control valve, consistant pressure and pressure memory were maintained and downstream surges eliminated.

Pressure compensating emitters, with little variation in pressure, are essential because drip irrigation is sensitive to elevation changes within the zone. Twenty feet of elevation changes within the lateral line results in a 8.7 PSI pressure change. If the system is operating at 20 PSI, it represents nearly a 50 percent change in pressure. Keep the pressure fluctuation within the zone under 10 percent using pressure compensating devices.

With new plantings, it's important that emission points are 3 to 6 inches inside the edge of the root ball. If the emitter is at the edge of the root ball or beyond, most of the water will bypass the root ball. The plant root ball is usually compacted soil, but the soil around it is generally loosened — not compacted — backfill. Because it is less dense this loose soil tends to drain most of the water away from the roots inside the compacted ball.

Try to match the varying plant water requirements within the zone by using flow rates from 0.5 gph to 24 gph, and single/multiple emission points for each plant. A 1-gallon shrub might have a 0.5 gph emitter; a 5-gallon shrub, two 0.5 gph emitters; and a 2-inch caliper tree, six 1 gph emitters. Flow rate is proportional to the water requirements of the various plants within a zone.

Several years ago, a local water purveyor discovered a drip system that had run non-stop over a long holiday weekend. The trouble occurred because the automatic control valve failed to close. A combination electric control valve and pressure reducing valve was used, but

This sloppy installation produced the kink in the distribution tubing, and lack of water killed the junipers.

would not close under such low flow conditions. Check with the manufacturer to ensure that the valve you use will close under low flow rates.

If possible, use a pressure reducing valve which is separate from the automatic control valve. The combination valves are expensive, and tend to "forget" the pressure at which they were intended to operate. For cost containment and ease of operation, use a preset (at 40, 45 or 50 psi) plastic pressure reducing valve in drip irrigation systems.

Many systems use emitters with capillary tubing to the emission point (with a bug cap) staked above the root ball, soil and mulch. This avoids potential for root intrusion and allows personnel to see the emitter in operation.

Emitters installed on the emission end of the capillary tubing above ground level provide a better grip for vandals to pull the drip equipment out of the ground. In cases where vandalism is a problem, install the emitter outlet 3 to 4 inches below the soil surface and tape a handful of pea gravel wrapped in filter fabric to the end of the capillary tubing. This creates a larger surface area for the water to move into the soil.

As root barriers are perfected and education continues more efficient, reduced maintenance subsurface irrigation will be available.

Self-cleaning emitters with a flush rate, if any, of less than 110 percent of the drip flow rate, and a pressure regulation range of 30 to 60 PSI are often preferred. This allows for greater elevation changes and flexibility in zoning the drip system. The minimum size drip tubing should be 3/4 inch to allow for future drip demands and changes. The price difference between 1/2 inch and 3/4 inch isn't worth its smaller size and limitations.

As with other forms of irrigation, drip should be zoned for varying plant needs, slopes and exposures. (See Chapter 8 for more information.)

INSTALLATION. Drip distribution tubing should always be installed with a minimum 4 to 6 inches of soil cover. Even so, many landscape projects contain drip tubing installed under the mulch or staked to the soil surface. After only one or two years the tubing is clearly visible above the mulch and the stakes are forced out of the ground by the freeze/thaw cycles — and sometimes by animals and people.

If drip is used in cultivated gardens, distribution tubing should be installed below the tilling depth. In mulched beds and non-cultivated areas, the capillary tube should be installed several inches below the soil surface to better protect the tubing. A common method of attaching the emitter to the drip distribution tubing is by using an emitter with a barbed fitting and pushing it into a hole in the tubing. A hole punch is used to place the hole in the tubing.

Some contractors think they can save money, however, by using a different hole punch than specified by the manufacturer of the drip emitter. The job was finished and turned over to the owner's maintenance contractor. Within a few weeks, drip emitters were blowing out of the tubing on a regular basis. The manufacturer's rep was called out to the job site and it was determined that the punched hole was too large for the barbed fitting on the emitter. The problem was solved by plugging the large holes and using the appropriate hole punch. Always use the hole punch recommended by the manufacturer.

We prefer to install multi-outlet (6 or more) emitters on a 1/2-inch pipe riser connected to ridged PVC in the non-freeze (soil) climates. In freezing climates, a resilient, flexible pipe is preferred. The main advantage of the resilient pipe is that it's easy to snake the pipe through plant beds.

It's important to keep all of the components clean during installation and to flush tubing before, during and after the components are installed. An automatic flush valve should be installed at every dead end in a drip irrigation line.

Drip tubing sizes vary from one manufacturer to another, unlike PVC and polyethylene pipes which have standard outside and inside diameters.

Don't place emitters close to the trunk or base of the plant, and don't wrap the capillary tubing around the trunk of the tree. This may strangle the plant and kill it.

Elimination of turf in narrow areas and the use of drip irrigation could prevent the cracking of this concrete roadway.

On slopes with plants, emitter outlets should be placed on the uphill side of the plant, and the distribution tubing should be routed along the contour of the slope rather then vertically up the slope.

MAINTENANCE. Drip is generally used on trees and shrubs installed in mulched planting beds or in native grass areas. In ground covers or flower beds low-pressure (30 PSI), 12-inch pop-up spray heads are common. A big concern with drip in these areas is the potential damage from cultivating, weeding and rototilling.

As with other types of irrigation, it's important to check the system weekly and the soil more frequently for system function and adjusted schedules. If a large number of emitters are failing, contact your distributor or the manufacturer for assistance in solving the problem.

In cold climates care should be taken when blowing out drip lines because the air pressure (don't exceed 25 PSI) can become too high, driving components apart.

When chloride and/or sodium are found in the water supply or soil, salt leaching is required to ensure plant health. When designing emitter locations, it may be necessary to provide 100 percent coverage of the wetted area over the active root zone in order to flush the salts away from the roots.

Once this is done, the salt builds up along the edges of the wetted area where evaporation and drying occur. This sometimes results in white rings on the soil surface. Rain can also force the salts back into the root zone. This can be prevented by turning on the system while it rains and shutting it off after a couple of inches of rain have fallen.

By going to subsurface drip irrigation, vandalism could be eliminated with all the components buried and accurately located on an as-built drawing. For information on drip irrigation scheduling, see the first four chapters on water management.

As contractors become more familiar and confident with drip, and as drip emitter outlets are protected from root intrusion and plugging, subsurface applications will be the common method.

Several subsurface drip lines are available that will resist root intrusion according to a three-year study at the Center for Irrigation Technology in Fresno. We are currently testing these products using subsurface drip lines in narrow areas for planting beds and turf where spray irrigation is wasteful. We have recommended some of these products to our clients, when appropriate.

QUESTIONS:

1. Should the pressure reducing valve be installed upstream or downstream of the electric control valve?

_____

2. Where should the emission point be placed for new plantings?

_____

3. How should drip distribution tubing be installed?

_____

4. Do most electric control valves close under low flow conditions (1 GPM or less)?

_____

5. What is the best way to provide uniform flow within a drip zone?

_____

6. Is filtration required if potable water is used for drip?

_____

7. What are the two most important factors when designing drip?

_____

8. Should pressure compensating emitters be used?

_____

9. What is the allowable pressure fluctuation within a drip zone?

_____

10. What are the benefits of using a preset plastic pressure reducing valve?

_____

(Answers to these questions are found on page 254.)

12

WATER MOVEMENT AND ELECTRICITY

MOST IRRIGATION SYSTEMS are controlled by electrically activated timers, valves and sensors. The wiring for these components must be properly sized and connected to ensure proper operation of the irrigation system. If the conductor size is too small, the voltage for the valves and sensors may be inadequate to operate the equipment. Conversely, if the voltage is too high the equipment may be damaged.

Water movement within piping systems is somewhat similar to the flow of electricity within wiring systems. Water pressure is like voltage or power. Accordingly, water volume and flow rate compares to electrical current in amperes, and water pressure loss equates to electrical resistance which is measured in ohms. Voltage is the force applied to the flow of electrical current and is measured in volts. It can be compared to a pump or static water pressure that supplies the irrigation system. This force is measured in volts (V or VAC for alternating current vs. DC for direct current).

Alternating current is provided by electric utilities for most of the irrigation controllers on the market, and direct current is used primarily by solar and battery-powered controllers.

Current is the flow of electricity and can be compared to the flow requirements (gpm) of a sprinkler head. Current is measured in amperes or amps (A). The solenoid on a control valve requires a higher amperage inrush current to actuate the solenoid. A lower voltage generally is required for a holding current to maintain the solenoid position during operation.

Electrical resistance, measured in ohms (R) is found in wire and other components. In water, friction loss causes a drop in pressure as it moves through the pipe. The same is true with electrical conductors carrying current. Voltage drop is also caused by resistance voltage which varies with wire length, diameter and the conductor material.

A review of Ohm's Law will aid in the understanding of voltage, amperage and voltage drop. This shows the relationship between voltage, current and resistance:

$$V = I \times R$$

Where:

    V = Voltage in volts
    R = Resistance on ohms
    I = Inrush current in Amps

and

    R = V/I
    I = V/R

This mainline break demonstrates the force of water pressure. In comparison, if the water were electricity then pressure would be voltage.

When the circuit resistance is constant, the voltage can be changed by varying the current. When the current is constant, the voltage can change by varying the wire size and the resistance.

Most equipment is designed to operate at plus or minus 10 percent of its rated voltage. A device rated at 120 volts has an operating range of 108 to 132 volts. A solenoid rated at 24 volts has an operating range of 21.6 to 26.3 volts.

But beware. The lower voltage requires more current or amperage to operate the equipment, and higher current produces more voltage loss in the conductor.

The power supply for the irrigation controller is usually obtained from the local electric utility which feeds 220-volt power though a meter to the circuit breaker box where it is split into two separate power supply legs of 120 volts each. In some areas of the country, this may only be 115 volts each.

These two 120-volt power source legs are out of phase with each other. All controller power supplies must be in phase with each other or they will cancel each other when connected to the same common valve neutral wire. However, this does not meet the National Electrical Code requirements.

Wires twisted together without UL approved connectors will eventually short to the ground, and the system will fail to operate.

Generally, every other breaker from top to bottom are in synch. If the voltage at output connectors of the circuit breaker read 220 VAC, that power source is out of phase. However, if it reads 0 VAC, the source is in phase and can be used.

The National Electrical Code and industry standards require a separate common valve neutral wire for each controller.

Good design practices limit the voltage drop to 2 percent of the rated voltage. A device rated at 120 volts could afford a voltage drop of 2.4 volts, and an operating range from 117.6 to 122.4 volts. Most controllers have an electrical input rating of 115 to 117 VAC at 1 amp or less, and an output of 24 to 27 VAC at 2 amps or less.

WIRES. Sizing the wire requires the following information: wire circuit length, allowable voltage loss in the circuit and inrush amperage. The allowable voltage drop is equal to the minimum voltage at the power source, less the voltage required by the equipment. Equipment voltage and amperage requirements are available from manufacturers.

$$V = \frac{CL \times I \times R}{1,000}$$

Where:

- V = Voltage in volts
- CL = Circuit length (double the distance between the clock and valve)
- I = Inrush current in Amps
- R = Resistance in ohms

For example, the wire length from the controller to the valve is 1,650 feet. Double the length of the wire to complete the circuit and it equals 3,300 feet. With a peak demand current of 0.30 amps and a resistance of 2.58 ohms per 1,000 feet of 14 AWG wire equals:

$$\text{Voltage drop} = \frac{3,300' \times 0.30 \text{ amps} \times 2.58 \text{ ohms}}{1,000} = 2.55 \text{ Volts}$$

A voltage loss of 2.55 is not within the allowable 2 percent loss of 2.4 volts for a 120-volt supply. At 115 volts, the allowable loss is 2.3 volts. To correct this problem the common wire size could be increased to 12 gauge and calculated as follows:

$$\text{Voltage drop in 1 leg (\#12)} = \frac{1,650 \times 0.30 \times 1.62}{1,000} = 0.80$$

$$\text{Voltage drop in 1 leg (\#14)} = \frac{1,650 \times 0.30 \times 2.58}{1,000} = 1.28$$

Total of both legs: 0.80 + 1.28 = 2.08 voltage drop, which meets the 2 percent standard.

Wire Gauge (AWG)	Wire Type (UF)	Resistance in Ohms per 1,000 feet 77F	149F
18	Solid	6.51	7.51
16	Solid	4.09	4.73
14	Solid	2.58	2.97
12	Solid	1.62	1.87
10	Solid	1.02	1.18
8	Solid	0.641	0.739
6	Stranded	0.403	0.465
4	Stranded	0.253	0.292
2	Stranded	0.159	0.184
0	Stranded	0.100	0.116

Note: Use 77F temperature for wire buried underground and 149F for above ground wiring. The resistance in copper wire increases with the rise in temperature.

Determine the maximum distance for a 12-gauge wire circuit using the following formula:

$$\text{Length} = \frac{\text{allowable voltage loss} \times 1,000}{\text{amp} \times \text{resistance in ohms} \times 2}$$

or

$$\text{Length} = \frac{2.3 \times 1,000}{0.25 \times 1.59 \times 2} = 2,893 \text{ feet}$$

If you want to avoid the formulas and do it the easy way, see the Wire Sizing chart in the Appendix for the allowable wire lengths for different sizes of wire and various amperage requirements.

High water pressure can effect the operation of the solenoids on some valves. If the water pressure at the valve is more than 80 PSI, check with the manufacturer to determine the additional voltage/amperage that may be required by the solenoid.

LIGHTNING. Lightning striking the ground can damage and/or destroy irrigation electrical control systems. Lightning and surge protection are necessary to protect these controls. A low-resistant earth ground is the first course of action followed by surge protection on the power supply side of the controller. In addition, surge protection should be provided for the 24-volt outlets at the controller. All 115 VAC irrigation controllers should be connected to an earth ground, whether installed inside or out. In many situations the earth grounding can be done by grounding back through the three-legged power supply.

Be careful. When controllers are a considerable distance from the power supply or pedestal mounted, they should be earth grounded at the controller location. This will prevent you from being the "ground" or liable for an electrocution. This is also required by law in the National Electrical Code.

A ground conducts electricity and provides a discharge path for short circuits, power surges and lightning. A ground is an electrical connection with the earth. The earth ground is usually copper wire from the controller to a copper rod in the ground, or a copper wire connected to a metallic water pipe that originates in the earth. A ground is measured by its resistance in ohms to the surrounding earth. Accepted industry standards require that grounding not exceed 15 ohms of resistance in the earth ground. If conditions permit, stay under 5 ohms.

If you can't meet these standards several steps may be taken including the use of longer rods and/or setting the rods deeper in the ground, using parallel rod systems or chemically raising the salt level of the soil. A good ground is usually less than 15 ohms and may vary in resistance depending on soil moisture, soil minerals and soil temperature changes. Soil moisture should be kept at a minimum of 20 percent.

Either water the ground rod area with an extra pop-up sprayhead, or install the rod in a drainage swale or low area to maintain the 20 percent moisture content. If the soil moisture content is zero, the soil is a perfect insulator.

Soil temperatures vary throughout the year. As the soil temperature decreases the soil resistance increases reducing the effectiveness of the earth ground. Below are the approximate levels of resistance in ohms for various soil temperatures.

Soil temperature F	Resistance ohms
68	7.2
50	9.9
32	13.0
32 (ice)	30.0

Because soil conditions may change or the grounding system can be damaged, check the system annually for resistance to determine if it still meets standards. This is particularly important in lightning prone areas.

Surge protection or suppressers should be installed on the power supply or input side of the controller. In areas where lightning is a problem, install surge suppressers on the 24-volt output lines. The surge suppresser can be grounded to earth for additional protection. Coil the wire at the solenoid (20 turns around a 1-inch pipe) to dissipate electrical surge from lightning. Wire should be buried 24 inches for 110-volt or higher voltage, and a minimum of 6 inches for circuits of 30 volts or less. See the National Electrical Code for additional information.

QUESTIONS:

1. Which wire has the smaller diameter, a #8 or #10?

_____

2. What is voltage?

_____

3. What is the industry standard for voltage drop?

_____

4. What is current?

_____

5. When should surge protection be installed on the 24-volt outlet of the controller?

_____

6. What is inrush amperage?

_____

7. Does temperature affect the resistance of the conductor?

_____

8. What percent soil moisture is required to maintain a good ground?

_____

(Answers to these questions are found on page 254.)

13

INSTALLING IRRIGATION SYSTEMS

CONTROL AND DELIVERY systems are vital elements of an irrigation scheme. Chapters 13 and 14 will give you some guidelines for the correct installation of an irrigation system. System components are divided into two major categories: control systems and delivery systems. Control systems are the topic for this chapter with discussion centered around valves, controllers and sensors.

Proper installation of system components is critical for a cost-efficient, easily maintained, water conserving, long-lasting system. Safety and reduced liability are also dependent on good installation practices.

Before beginning an installation job, check the static and operating water pressure, flow rate and the size of the water supply to verify that design conditions are the same as actual site conditions.

Many of Keesen Water Management's installation detail drawings are located in the Appendix beginning on page 225. Please refer to these as you are reading Chapters 13 and 14.

VALVES. Install the zone control valve with the flow control completely open or turned down one or two turns (if required by the valve manufacturer), and the top of the flow control 4 inches below the finished grade. This allows the valve box lid to clear the valves while providing good access for maintenance work. Place a resilient seated gate valve upstream of the zone control valve for easy repair and maintenance of the control valve as well as for emergency shut-off. If quick couplers are needed, place these within the valve box, upstream of the isolation valve so they can be easily located in the future.

Place 4 to 6 inches of clean, washed gravel under the valve boxes for good drainage. Install filter fabric under the gravel and attach it to the valve box with duct tape. This fabric keeps the soil from working its way up through the gravel, or silting in along the pipe and forming a mud hole. Install the valve boxes flush with the finished grade and lightly compact the adjacent soil

When a valve is installed too high in the valve box, the lid won't properly fit creating the potential for liability to pedestrians and damage to equipment.

to prevent settlement. Always install sturdy valve boxes to withstand pressure from mowing and maintenance equipment.

Many rectangular valve box lids are difficult to remove and replace because the two longest walls cave in when excessive pressure is placed against the wall. A piece of rigid PVC pipe can be placed within the valve box between these two walls to brace them and prevent a cave-in.

Never install valves in low areas where water can collect, or close to driveways or sidewalks where automobile damage and pedestrian safety is a concern. Brand all valve box lids with the control valve number or other appropriate identifiers such as MD for manual drain or IV for isolation valve. Plastic identification tags can be attached to the valve in case the lids get switched. Most valve box lids can be bolted in place to help prevent vandalism. When vandalism is a problem the valve boxes can be buried with 6 to 8 inches of soil. Use a brass valve or a metal foil to allow locating with a metal detector.

Coil 2 to 3 feet of control wire around a 1-inch diameter pipe prior to connecting it to the solenoid to allow for valve bonnet removal, and to protect the solenoid from potential electrical surges.

For better performance, the drip

A look inside a controller in which a contractor incorrectly substituted masking tape for wire nuts to make the connections to the controller.

emitter valve assembly should be installed as follows: Install the filter first to keep the rest of the components clean. The filter should have a pressure rating of 125 to 150 PSI. Install the strainer with an easy access flush valve on the top. Next install a fixed-rate pressure reducing valve upstream from the zone control valve. If installed downstream, pressure surges can occur within the space of time that the regulator takes to set itself after the water is turned on. Surges can blow emitters out of the lines as well as blow apart pipe and fittings.

A 1/4-inch outlet tee with an air valve placed downstream of the control valve is helpful to diagnose future system problems using a pressure gauge.

Isolation valves are necessary in larger systems to shut down portions for repair without shutting off the entire system. Isolating portions of the system will make pressure testing for leaks and leak detection much easier.

Quick coupler valves should be securely attached to 2-inch x 2-inch x 30-inch treated wood stakes, or rebar driven into the soil. This will support the valve when the coupler key is inserted, and prevent damage to the pipe and fittings.

This lid floats on the water to the left of the valve box in a low area making this area hazardous as well as a landscaping eyesore.

Any valve, other than the control valves, installed below grade should be installed in a 4- to 6-inch diameter PVC pipe access sleeve for valve control. Pipe should be cut out on one end to better fit over the pipe/valve and prevent the sleeve from shifting away from the valve. Center the sleeve on the valve operator and set it vertically with a 10-inch round valve box at the surface. The larger size sleeve makes it much easier to vacuum or pick stones out of the system.

Backflow preventers should be placed inside buildings or in shrub beds screened with planting whenever possible. If the backflow preventer is installed in a turf area, pour a 4-inch thick concrete pad that is 12 inches greater than the backflow and/or preventer enclosure to act as a mowing strip.

At Keesen Water Management, backflow preventer enclosures are included on every design project except single family homes. They prevent people from shutting off valves, opening petcocks and breaking valve handles by hiding the temptation from view.

For protection, backflow preventers close to driveways or streets should have 4-inch or larger diameter galvanized pipe filled and set 30-inches deep in concrete with a diameter of 24-inches. Insulated enclosures will protect the backflow preventer from freeze damage in the early spring and late fall.

Most flow sensors and meters require a minimum straight pipe length equal to 10 times the nominal diameter of the pipe on the upstream side, and up to 5 times diameter on the downstream side. If valves or fittings are too close to the sensor, the accuracy will lessen due to the turbulence in the pipe.

An air valve should be installed downstream of every pressure reducing valve to aid in adjusting the PRV and checking operating pressure.

Root intrusion into this valve box could have been avoided if the valve wasn't installed in the drainage swale, and if filter fabric had been installed under gravel, below the valve and taped to the outside of the valve box.

CONTROLLERS. Be safe. Always ground the controller by using an Underwriters Laboratory (UL) approved 5/8-inch x 8-foot copper grounding rod and proper ground clamps. Each controller should have its own ground rod. Install a protective conduit around wire exposed above ground level. This will protect the wiring as well as provide a better looking installation. Controllers and electrical equipment need to be protected from vandals and snow removal equipment, but installed in a manner that can be easily accessed by maintenance crews.

To save time on the installation and maintenance of an irrigation system, include an outlet plug for a hand-held radio remote control device. This allows easy operation and testing of the system anywhere on the landscape site or location.

Mounted on the top of the building, where the freeze sensor and rain shut-off sensor cannot be vandalized, these sensors override the controller when low temperatures and rainfall occurs.

Pedestal-mounted controllers should be installed on a concrete pad similar to the pad for the backflow preventer. Two sweep elbows should be used, one for the 24-volt wires and another for the 120-volt wiring.

Avoid the installation of controllers and other electrical equipment below grade, unless the vault is well-drained and well-ventilated with fans. The high humidity that develops in vaults can cause corrosion and greatly reduce the life span of the electrical equipment.

Controllers that are installed outdoors should have watertight enclosures except in desert climates where ventilation and cooling of the controllers are more important. In addition, avoid installing the controller close to an irrigation head and, if possible, keep it out of exposed turfgrass areas.

Controller charts are reduced size, as-built drawings containing zone numbers with the zone coverage areas highlighted in different colors, laminated in 20 mil plastic and mounted in the controller door. This is helpful in identifying problems and when trying to determine which zone serves what part of the landscape.

All wire connections made below grade should be UL approved, removable insulated wire nuts installed in a reusable, watertight plastic container filled with a gel and installed in a valve box for future access.

Wire should not be pulled through the ground as this may cause the wire and insulation to stretch and eventually break. Wire can be laid with a cable plow or installed in an open trench. Tape and bundle the wire at 15-foot intervals when installing in a trench. If the wire is installed with the mainline, place it to one side and several inches below the top of the pipe. This will help protect the wire from damage that may occur from future excavation.

Provide a 24-inch expansion loop for wire whenever a change of direction is greater than 40 degrees as well as in situations where the length exceeds 300 feet. Wire that is tightly stretched in a trench may separate within its insulation as soil temperatures cool down.

SENSORS. Some sensors are sensitive to the effects of electromagnetic fields and require shielded cable and connectors between the sensor and the controller.

The 24-volt wiring should be buried at least 12 to 18 inches deep to better protect the wire, and 120-volt wiring should be buried 24-inches deep according to the national electrical code. All wire should be UL approved for direct burial.

Several years ago, we did some consulting for an irrigation contractor in Kentucky. We recommended the use of a rain shutoff device for his systems, but several months later he called to say the rain sensor was not working correctly and the turf was burning up. My schedule placed me in Kentucky that week so I arranged to look at the installation with the contractor.

The rain sensor was mounted on an 8-foot high fence in the back yard. When the system came on, the rotor heads hit the sensor and shutoff the system. Rain sensors are best installed above the spray height of the system, away from trees. A good location is at the roof line of a building.

QUESTIONS:

1. Why is the fabric installed under the valve box?

_____

2. What degree of compaction is required around valve boxes?

_____

3. Is a 2-inch sleeve adequate for an isolation valve?

_____

4. How are flow sensors installed?

_____

5. What should be installed downstream of every pressure reducing valve?

_____

6. Should wire be pulled into the ground with a pipe puller?

_____

7. What happens when electrical controllers are installed below grade?

_____

8. What should be installed downstream of every pressure reducing valve?

_____

9. What are controller charts?

_____

(Answers to these questions are found on page 255.)

14

PROPER INSTALLATION GOES A LONG WAY

SAFETY AND LIABILITY exposure should be top concerns for irrigation contractors. Unfortunately, many contractors are installing systems in a sloppy manner without regard for safety, quality, longevity, ease of maintenance and efficiency.

Property owners and citizens have sued and won damages from irrigation contractors for inadequately installed systems. There's no need, however, for the situation to reach this critical juncture when it can be avoided by properly installing irrigation systems. Adhering to the guidelines detailed in this book should help eliminate potential troubles down the road.

INSTALLATION DEPTHS. The biggest problem is setting heads and valve boxes to the finished grade. Millions of dollars a year are spent on irrigation system repairs due to mower damage. Most of this unnecessary damage results from the heads and valves being set too high in the ground.

Initially, most damage to irrigation equipment is the liability of the irrigation contractor. Later, the maintenance contractor becomes liable as he manages the system and is the primary person responsible for proper repair and maintenance.

Because repaired parts are reinstalled at the exact same height as the original, repeat damage often occurs to these heads and valves as well. The finished grade in the turf area should be the surface at the base of the grass blades. The top of the head should be rest about 1/8 inch below this level to avoid interference with mowers and pedestrians.

Sprinkler heads should always retract below grade, while spray heads in shrubs and ground covers should be 6- or 12-inch pop-up heads placed out of the way of pedestrian traffic and snow plows. Allow at least 2 inches, or three finger widths, between the heads and the edge of a driveway or sidewalk where turf edging equipment will be used. Allow 6 inches in planting beds.

Place heads perpendicular to the finished grade (except on slopes greater than 3:1. See

Chapter 7 on Sprinkler Head Selection and Placement). Always hand tamp soil firmly around the head to prevent movement and erosion and, when possible, install the head against undisturbed soil for greater stability.

Every time a zone is turned on there is some surge in the lateral line. This surge potential is greater in systems without check valves in the heads because water will drain out of the lateral after every operation.

For surge protection and water conservation, Keesen Water Management designs irrigation systems with check valves installed in the base of the sprinkler head. Lateral line surge can cause the heads to move in the soil and fittings to blow apart. Good compaction around the sprinkler head will reduce the effects of surges, and will keep the head in place.

PVC pipe installed at the surface will damage quickly from mowers and aerators. Proper installation will protect the pipe and reduce maintenance costs.

MAKING CONNECTIONS. If force is applied to the head the connection between the sprinkler head and pipe should be flexible enough to shift without breaking or weakening pipes and fittings. Fittings should be durable and stronger than the pipe to which it connects.

Threaded connections are the weakest type of connection in the irrigation system. Use solvent welded joints whenever possible with PVC pipe. Use Teflon tape on all male threads to prevent leakage. Avoid oil-based compounds because they can weaken the PVC and cause leaks.

Fine tuning a recently installed irrigation system reduces waste, improves the system efficiency and places the water where the plants can use it.

In my early years as a contractor, we started the season using a new pipe thread sealant that was easy to use. Three months later we were replacing valve manifolds on virtually every job we installed. Needless to say, we didn't make a lot of money that year.

A 14-inch to 24-inch length of highly resilient pipe (pressure loss in 1/2-inch pipe by 24-inches long is 0.127 PSI and a velocity under 5 fps), that will not kink when a 12-inch length is formed into an arc and connected between the head and the pipe seems to be the best method for heads with 1/2-inch and 3/4-inch connections.

Larger heads with higher flows require PVC swing joints. Several years ago we discontinued specifying the double swing joint made up of threaded street ells and nipples, and are now using a manufactured double-swing joint with "O-Ring" seals that provide flexibility without leaking.

This head was installed against the walk with the lip on top. Now the head will likely be damaged by a turf edger or snow plow.

On commercial projects avoid using PVC nipples or poly cutoff nipples as the only connection between the pipe and head. Properly designed small residential systems with poly laterals can use a PVC nipple connected directly to an insert tee or pipe saddle, because of generally smaller mowing equipment.

For commercial projects, always install heads on flexible tubing or on manufactured swing joints to protect lateral pipe and head connections from damage by mowing equipment or other vehicular traffic.

MAINLINE. Situations resulting in high pressure create the potential for surge (water hammer) which is caused by fast closing automatic control valves. Subseqeuntly, install the mainline with more cover than the lateral lines.

Install the mainline pipe at a depth of 18 inches if the pipe is 4 inches or less, and at a 24-inch depth for larger pipes. Residential systems are smaller so the mainlines can be installed at a 12-inch depth. This provides adequate cover to keep the mainlines in place, and avoids interference with the lateral lines which cross above the mainline.

Maintain an even trench bottom that continuously supports the pipe on a uniform grade. If two or more pipes are installed in the same trench, provide a minimum of 4 inches between the pipes to allow room for repairs. Use clean, rock free, backfill material around and above (6 to 12 inches) the pipe to prevent damage.

Compact trench backfill to a density equal to the adjacent undisturbed soil. Flooding or "puddling" the trenches is a good method to achieve proper compaction along with light tamping when the soil dries out. Avoid compaction with trucks and large tractors.

Install wires on the side of the mainline pipe, and tape and bundle wire every 15 feet to help prevent wire damage should the area need to be excavated.

Provide 24-inch expansion loops whenever wire direction change is greater than 40 degrees to allow for the expansion and contraction of the soil during temperature changes. Install several spare wires to avoid future problems.

Lateral lines can be installed 8 inches to 16 inches deep. The depth of turf aeration determines how deep to place the pipe. Eight inches of soil is more than sufficient to hold a 2-inch pipe in position. Pulling pipe

The 4-inch pop-up spray head pictured here was installed too low and becomes useless. Installing the head just below finished grade ensures many years of effective operation before the head has to be raised.

is acceptable if the soil is free of sharp rocks or other conditions that may damage the pipe.

For polyethylene piping, use insert fittings with screw clamps or locking pinch clamps. Double clamp 2-inch and 1 1/2-inch pipes, and make sure the clamp is installed over the serrated edges of the insert fitting.

Once the mainline pipe is installed, and before the valves are installed, backfill the pipe — except for all fitting connections. Next, flush any debris from the line and install the control valves. Check all of the connections once the mainline is pressurized, and then backfill.

The dark line in the turf on this side of the valve box is a hole in the turf where the mainline was not backfilled and compacted properly.

Another method for pressure testing uses a high pressure force pump which builds the system pressure up to 100 to 150 psi. If the pressure drops rapidly once the pump is shutoff, it's an indication that a valve is open or a leak is occurring.

After the installation is complete, check the head coverage for distance, uniformity and alignment. Adjust radii and arcs to improve system performance.

When routing the pipe, stay outside of the drip line of any tree as much as possible. Two years ago, one of our renovation designs was installed in a large office site with many parking lot islands and existing mature trees. Today, many trees have died because the roots were cut and the trees were without water during the growing season.

Flush the lateral lines before installing the heads. Prior to installing the last heads on the ends of the lateral lines, flush the line again to remove the soil that entered the line while installing heads.

As-built conditions should be noted on the irrigation plans as the work progresses and transferred to a reproducible drawing at completion. The as-built plan should show the dimensioned locations of the following:

- Point of connection
- Mainline location (every 100 feet)
- Isolation valves
- Electric control valves
- Quick coupling valves
- Control wire routing
- Sleeves

Dimensions should be recorded from two permanent reference points such as buildings and streets. The irrigation legend should be changed if the equipment installed differs from the original plans. An operating and maintenance manual should be provided for every project, including the name, address and phone number of the designer, contractor, local distributor and other pertinent details.

The beginning and duration of the warranty period is included along with an equipment list with the manufacturer's name, model or part number and detailed operating instructions and maintenance guidelines.

Using the ideas in this chapter will help you install a more durable and efficient irrigation system and will help avoid future liability problems.

QUESTIONS:

1. What is the biggest problem with many irrigation systems?

_____

2. What degree of compaction is required around sprinkler heads?

_____

3. At what height should sprinkler heads be installed?

_____

4. Will check valves in the base of the head reduce the surge potential in lateral lines?

_____

5. What is the weakest type of connection in the irrigation system?

_____

6. Should pipe be pulled into the ground with a pipe puller?

_____

7. How should heads be installed adjacent to the edge of a driveway or sidewalk?

_____

8. How can sprinkler head movement and erosion be prevented?

_____

9. How should clamps be installed on polyethylene piping with insert fittings?

_____

10. What is the best method of connection between the head (1/2-inch and 3/4-inch head inlet) and the lateral piping?

_____

(Answers to these questions are found on page 255.)

15

THE RIGHT STUFF

A LITTLE PREVENTIVE maintenance goes a long way when maintaining irrigation systems in the landscape. Most property owners don't realize the importance of preventive maintenance for the irrigation system. They only react when a dry spot appears in the landscape or a leak causes damage. The irrigation system is like any other mechanical device. It requires periodic maintenance to maintain proper performance and longevity.

Irrigation systems should be checked periodically for various adjustments, cleaning and functioning. The frequency is determined by the quality of the equipment and the installation. If the equipment is not installed properly, mowing equipment may cause repeated damage to the heads and valve boxes.

Older (more than 10 years) systems and poorly installed systems may require weekly maintenance, after every mowing. Newer systems can be inspected biweekly, and quality systems can be checked every three or four weeks.

Recommended irrigation inspections and how often they're needed:

Weekly to monthly intervals:

- Arc and radius coverage adjustments
- Proper head alignment
- Damaged equipment
- Potential liability
- Leakage
- Cleaning system filters
- Seeping control valves
- Slow closing valves

Annual maintenance:

- Lubricate pump
- Test backflow preventers
- Review system operating pressure
- Nozzle wear if sediment enters the system
- Test controller backup battery

- Flush drip filters
- Winterization where applicable
- System activation where applicable

Irrigation systems with good preventive maintenance programs last longer, perform better and provide for a healthier landscape.

WINTERIZATION. If the system is installed in a region subject to winter freezes, it's necessary to protect the equipment from possible damage. Irrigation in freezing climates has always been a challenge for designers, contractors and maintenance personnel.

This irrigation system is being "blown out" with compressed air after a late October snowstorm.

When water freezes in an irrigation system it expands enough to break the pipes, valves and heads. Most freeze breaks occur in backflow prevention devices and PVC pipes resulting in expensive repair bills, and sometimes flooded buildings and damaged structures with tremendous liability for all involved.

To complicate matters, a freeze/thaw cycle can occur in some winter climates. This enables water to re-enter the system via the irrigation heads and cause freeze damage even if the system is drained.

EQUIPMENT. Protect an irrigation system from damage by using freeze resistant equipment that allows for removal of water from the system.

First consider the pipe. There are two types of pipe commonly used in landscape irrigation systems. The first, poly vinyl chloride (PVC), is a semi-rigid, high pressure (160 psi and higher) pipe. The second type is polyethylene (Poly) which is a flexible low pressure (between 80 and 100 psi) pipe.

In most cases, use PVC for the system mainline. PVC will withstand the higher water pressure and the potential water hammer found in many mainlines. For residential irrigation system installations, however, high quality Poly pipe with a pressure rating of 125 psi or more is acceptable for short mainlines.

Poly is the best choice for lateral lines because it is flexible and more resilient. This is true only if it is NSF (National Sanitation Foundation approved). NSF pipe has better quality control during manufacturing as well as heavier wall thickness. This prevents major freeze damage even if the pipe is full of water all winter.

Install mainline pipe at a depth of 18 to 24 inches if the pipe size is 4 inches or less. Lateral lines are located at a more shallow depth. Pipe installed any deeper makes maintenance more difficult. It is senseless to bury the pipe below frost level because all the risers to the surface will freeze. Additionally, frost levels vary from year to year.

Where have all the nozzles gone? Weekly preventive maintenance could have averted this problem.

Plastic, automatic drain valves with a resilient seal closing at about 10 PSI of pressure can be used, but are not recommended. Brass and stainless steel automatic drain valves are not as effective because they corrode rapidly from the constant exposure to moist air during the winter. Automatic drain valves can be used, but every time the system is operated, water in the lateral lines will drain out of the pipe. This results in wasted water, surge damage and potential wet spots in the landscape.

A self-draining system using a plastic automatic drain on mainlines is feasible unless the mainline or backflow preventer is more than 20 feet above the automatic drain valve. With more than 20 feet of head, the water pressure is in excess of 8 psi at the drain valve, preventing the drain valve from opening. This drain valve is considered ideal for residential systems.

For best results, manual drain valves should be installed on all mainlines at low points, even if the system is blown out with compressed air. Manual drain valves should have a rising stem, rubber seat, brass body and angle configuration. A rising stem valve keeps the valve open when there is no water in the line. Drain valves vary in size from 3/4-inch to 2 inches depending on the size and length of the line being drained, as well as the time allowed for draining.

Use 3/4-inch to 1 1/2 inches of clean gravel for the drain sump. Wrap the gravel and valve box with a filter fabric (see detail) to prevent soil from entering the system. Drain sump sizes will range from 1 to 9 cubic feet depending on the size of the drain valve and the soil type. Use a small drain sump in sandy soil and a larger one in heavy clay soil. Drain valves can be drained to daylight by piping from the drain valve to a point above grade.

Replacing a broken head in this manner will guarantee damage. It could also cause someone to trip and fall ending in a lawsuit and lost client.

Designers and contractors should provide written winterization instructions to the owner or maintenance contractor. In reality, however, when the as-built plans are furnished 95 percent of them always seem to get lost because of the turnover in property owners and maintenance personnel. Solve this problem by installing as-built irrigation plans that are reduced in size and laminated for installation in the controller door. This will at least help the maintenance personnel locate equipment and maintain the system. Good "as-builts" are important.

WATER REMOVAL. You can remove the water from the irrigation system by using one of the following techniques:

The original method called for the installation of manual drain valves at all low points in the system. This was modified in the 1950s to include the use of brass ball check automatic drains on lateral lines only.

Systems with drain valves are more expensive for initial construction, and require pipe installation with at least a 1 percent slope toward all low drainage points. Manual drain valve boxes should be at least 10 inches in diameter for easy spotting. Systems with drain valves allow water re-entering the system to pass through.

This pop-up irrigation head is spraying water onto the fence because it was installed facing the wrong direction.

A second technique is to blow the water out of the system with compressed air. This method was originally developed because "as-built" drawings have a tendency to disappear. As a result, maintenance personnel couldn't locate manual drain valves or feel confident that they found all the drains. This method is now most common, and is necessary when sprinkler heads are installed with check valves in the base of the head.

Blowing out the system with compressed air is not foolproof and can leave water in the lines. But it has proven effective when done properly. In some situations — such as a drainage swale where piping dips through a low point — the air can blow the water out of the top part of the pipe and leave water in the bottom to freeze. This can weaken the pipe and cause a future break.

To reduce the potential for damage, never exceed 100 PSI of pressure at the air compressor and allow the air to enter the system slowly, allowing the pressure to build up gradually and avoid high surge pressure. "Blowing out" the system with compressed air is a more expensive procedure than using drain valves.

At Keesen Water Management, irrigation systems are designed to use a combination of the two methods. The system is designed for blowing out the water with compressed air using a quick coupler valve installed immediately downstream of the backflow preventer as the point of connection for the compressor.

A word of caution: Air is 1/5th the density of water and very compressible. Large systems require portions of the system to be isolated and blown out separately to avoid the use of higher pressures during water evacuation.

Compressed air can be hot when it emerges from the compressor. At least 100 feet of hose is recommended between the compressor and the system to allow the heat to dissipate. Manual drains are also used at low points in the system, along with poly lateral lines without automatic drain valves. Insulated enclosures are also used to protect the backflow preventers from freezing in early spring and late fall.

When installing 12-inch pop-ups, always use the bottom inlet to connect the head to the lateral pipe. This will help prevent water freeze damage to the pop-up head body and avoid debris accumulation in the bottom of the head.

Keep electricity running to the automatic controller on all winter to provide some heat and to protect sensitive parts of the controller.

TYPICAL SHUTOFF INSTRUCTIONS.

1. Close water supply valve(s).
2. Connect the compressor to the system.
3. Start pumping compressed air into the system only after the first station on the controller is activated or opened. If no valves are open, severe damage can result from the high air pressure in the mainline. At the automatic controller, cycle through each station allowing each valve to open for approximately five minutes or until all the water is evacuated from each zone.
4. Open manual drain valves.
5. Open petcocks and drain valves on backflow preventers.
6. Rest easy the entire winter.

The damage to irrigation systems in freezing climates can be removed by following these directions and designing for freeze conditions.

QUESTIONS:

1. Where do most freeze breaks occur?

_____

2. At what frequency should the irrigation system be checked?

_____

3. Can water re-enter the irrigation system during the winter?

_____

4. What type of pipe is the most resistant to freeze damage?

_____

5. How often should a quality, properly installed system be checked?

_____

6. Can blowing out the system damage the equipment?

_____

7. Name at least three items that should be checked regularly during the irrigation season?

_____

8. How often should older (more than 10 years) systems and poorly installed systems be checked?

_____

(Answers to these questions are found on page 255.)

16

THE HIDDEN MENACE

SOMETIMES AN IRRIGATION glitch is not easily recognizable making it necessary for the maintenance contractor to perform troubleshooting measures by trial and error to arrive at a solution.

Proper repair and maintenance of irrigation systems is critical for efficient operation and longevity. Systems that lack continued maintenance will undoubtedly have to be renovated or replaced before the end of its life cycle.

Irrigation systems are similar to automobiles or heating/cooling systems, requiring regular maintenance by competent, knowledgeable and well-trained individuals. It's sad to say, however, that the irrigation maintenance industry lacks properly trained individuals. Many system maintainers think they know it all, when in reality they have much to learn.

As you read this book, make a mental note of the items not currently being performed by your firm or those items that are being executed incorrectly. (Refer to the previous chapter on preventive maintenance for check lists.) Be honest. Are you really following these maintenance guidelines, or could you provide a better service for the owner of the irrigation system?

Depending on the age of the system — how well it was designed, installed and whether it has been properly maintained — the system should be operated and every head inspected on weekly or monthly intervals.

IRRIGATION TROUBLE-SHOOTING. Breaks in the irrigation system allow soil to enter the structure and cause plugged valves and heads. City water supplies can often transfer rust and soil into the irrigation system as well as non-potable water quantities. Likewise, if the water has a high mineral content, deposits can build up restricting the flow of water and leading to clogged nozzles.

As the flow is restricted pressure will increase, uniformity will deteriorate and water will be wasted. If the wiper seal on a head is leaking, operate the head and gently step on the top of the nozzle to force it down a few times. This may dislodge debris around the wiper seal and within the nozzle. If the seal is still leaking, remove the top of the head from the body, clean

out debris and replace the seal if it's damaged.

Suspended solids in the water can cause abrasive action which can enlarge the nozzle resulting in reduced operating pressure and radius, and poor uniformity of coverage. An inconsistent pattern may appear as a dry donut within the configuration. Use a set of drill bits to verify the nozzle size. Install a new, original equipment, correct-size nozzle if the current nozzle is too large.

Be careful when cleaning nozzles. Remove the nozzle from the head before cleaning and flush the head with a jet of water to remove any particles. At the same time, force foreign particles from the nozzle by spraying in both directions. If the nozzle is still clogged, take a stir stick or other small plastic device to pick out the object. If it's still plugged, replace the nozzle with a new one of the same origin. Operate the head again and check for proper radius, arc of coverage and performance.

Using a knife, wire or screwdriver to clean a nozzle may score the inside of the nozzle surface, disturb the distribution pattern and ruin the nozzle. When replacing heads be sure to match the performance to the previous head or to existing zone spacing and precipitation rates. Don't remove sod from around the head to help the spray trajectory. Not only is sod removal a real landscape eyesore, but the heads will plug more easily if they are sitting in a mud puddle. Raise the heads or install new heads with a higher pop-up height.

If a replacement head is longer than the original head, adjust the fittings under the head to allow the top of the head to be set 1/8-inch below the surface. If backfill soil is wet and unstable, remove and replace with dry granular soil and compact. Whenever a head is installed or replaced, the soil backfilled adjacent to the head should be compacted with a blunt instrument or by stepping on the soil. Heads not properly compacted, will tilt and turn when the mower wheel rides over the head.

VALVES. Slow-acting electric control valves won't open or close and may have a plugged orifice blocking the flow of water to and from the top of the diaphragm. To clean, remove the valve bonnet and use a small copper wire to clean the ports in the bonnet and body of the valve. If possible, rinse with water, reassemble and test.

Here two heads are causing problems in the landscape. One head is broken and the second is set too low in the ground.

If the valve won't open, check the flow control to see if it is open. Next try the manual bleed valve. If this operates the valve, check the solenoid and the controller for power with a volt meter or a good solenoid. If a clicking noise is heard at the solenoid when power is applied, the solenoid is probably working. If no noise is heard, the solenoid should be replaced. If the solenoid is good and voltage is sent from the controller, the system most likely has a problem with the wiring somewhere between the controller and the valve.

Seeping valves are usually caused by particles imbedded in the rubber seat of the diaphragm or lodged between the diaphragm and valve body. Over time the diaphragm may stretch out or develop cracks which will cause it to seep and eventually stay in the open position. A weak spring can also be the culprit.

REPAIR OR REPLACE. The decision of whether to repair or replace equipment is dependent on costs, adequate performance and repairability. For instance:

1. Replacement is necessary if the irrigation system components can't be repaired due to unavailability of parts.

2. The age and performance of the equipment may require replacement. Older heads that don't pop-up, and early model pop-up heads with a 1- to 2-inch pop-up height are often blocked by the height of the turf and soil buildup. This results in blocked trajectory spray which causes wet and dry spots and runoff. Replace it with a head that will pop-up 4 inches or more depending on the rate of soil buildup and the height of the turf prior to mowing.

   An example of age and quality was quite evident when Keesen Water Management evaluated the irrigation system for a medical facility and found several leaks in the lateral lines. When repaired, another leak occurred within the same zone.

   If the lateral line leaks in a different location every time repairs are made, then the pipe may be defective or the pressure could be too high. We discovered that the pipe was splitting because of a manufacturer's defect. The only solution was to replace all of the pipe. If the pressure is too high, install a pressure reducing valve to control it.

3. Repairing and/or replacing equipment usually depends on the cost of repair parts and labor vs. the cost of new equipment. The greater the cost of the equipment, the more likely it will be repaired. A well-trained, experienced service person is worth a minimum of $15 to $20 per hour — costs which may bill out at $30 to $40 per hour or more.

Using an hourly rate of $30 per hour, compare repair vs. replacement costs. A new 4-inch pop-up spray head with check valve and plastic nozzle costs about $3 plus 15 minutes ($7.50) to replace for a total of $10.50, plus markup on materials and travel time. Repairing a 4-inch pop-up with a broken riser stem may cost as much or more than the cost of a new head

Don't remove sod from around the head because the heads are more likely to malfunction if they are sitting in a mud puddle.

($3) and 10 minutes labor ($5) for a total of $8.50 compared to $10.50 for replacement.

Don't buy repair parts for inexpensive heads, buy new heads and use them for parts. If you're not sure what's wrong with the head, pull the pop-up assembly out of the body of the head and insert a new one.

Small- to medium-radius internal drive rotors cost about $15 new and 15 minutes to replace ($7.50) for a total of $22.50. These types of rotors are often more difficult to repair, may take 30 minutes or more to repair the assembly and cost more than the price of a new head for a total of $30. Subsequently, it's wise to buy a new head and insert the assembly into the old body. The same size impact rotors cost less and are usually easier to repair.

Repair large radius rotors as well as valves, if possible, because of the increased equipment cost and labor. This is particularly true when replacing a valve. Electric 1-inch control valves range from $15 to $35 and it may be more cost-effective to cannibalize new valves for parts. Valves running 1 1/2 to 2 inches cost from $45 to $70, and may take several hours to replace costing a minimum of $75. Buy the parts and make the repairs to save on costs.

The next chapter will further discuss repair and maintenance issues such as verifying pressure, pressure reducing valve functions, pump maintenance and more troubleshooting and repair tips.

Questions:

1. What kind of device should be used to remove objects from nozzles?

_____

2. Can city water supplies plug irrigation equipment?

_____

3. What causes nozzles to become oversized?

_____

4. When replacing heads what performance characteristics are important?

_____

5. What's causing the problem if the valve won't open, the flow control is open and the manual bleed valve opens the valve?

_____

6. Should 1- to 2-inch pop-up heads be replaced?

_____

7. How can a plugged orifice blocking the flow of water to and from the top of the diaphragm be cleaned?

_____

8. What can cause valves to seep?

_____

9. What is the problem if the lateral line leaks in a different location every time repairs are made?

_____

10. What is usually the determining factor when deciding whether to repair or replace equipment?

_____

(Answers to these questions are found on page 256.)

17

DIAGNOSING SYSTEM ILLS

CONTROLLING PRESSURE, SOLVING hydraulic and electrical problems and pump maintenance seem to be the least understood topics in the irrigation design, maintenance and contracting industry. The information presented in this book is not all inclusive, but will get you started in the right direction. If you have additional ideas, please let me know.

Proper irrigation system operating pressure is of paramount importance if an operator wants to control costs. Subsequently, system operating pressure must be checked annually. If a pump is used, it should be checked at least monthly.

Visible changes in water pressure may appear in potable water supplies and pump systems because of increased area demand, equipment deterioration and/or pressure changes made by water purveyors.

Once it's determined the water pressure at the source is the same as the pressure specified in the design, obtain operating pressure readings at several points in the system. Turn on station number one and test the operating pressure at the pump or at the down stream side of the pressure reducing valve and/or the backflow preventer. Also test the pressure at the electric zone control valve or at one of the sprinkler heads within the zone.

Log the time, date, pressure and location for future reference. Do this test for at least the largest and farthest zone in the system. Also note the static water pressure when the system is off.

An easy way to check the operating pressure is to install schrader valves downstream of the pressure reducing valve as well as each zone control valve. Attach a quick-connect device with hose and gauge and read the pressure. The hose should be long enough to observe the gauge and not get wet from the sprinklers. Other methods of checking operating pressure can be performed at the head using a pitot tube/pressure gauge for rotor heads and a tee/pressure gauge inserted under the pop-up spray head nozzle.

Once the pressure data is logged it can provide a reference for proper operating pressure

The lateral line under this step leaked for many years, causing the concrete to erode until a hole appeared in the step. If this leak had been identified earlier, the owner would have saved a lot of money.

within the system. Periodic pressure checks can reveal if the pipelines are leaking and whether pressure reducing valves, backflow preventers, control valves, sprinkler heads and pumps are operating properly. System pressure variances can pinpoint problems as well as save time and labor.

If the source pressure remains unchanged and the pressure within the zone has changed more than 5 percent to 10 percent, try to isolate the problem with additional pressure tests throughout the system. Lower operating pressure at the zone might indicate a leak, worn pump impeller or a partially closed flow control at the zone valve. Higher operating pressure could be the indication of a plugged nozzle or pipe, a defective pressure reducing valve or an increased pressure from the water purveyor.

The most troublesome problems can occur between the irrigation system point of connection and the city water main. Two systems designed by our firm experienced low water pressure when they were installed. We monitored the construction and performed a final inspection for both, but something seemed wrong.

At the Federal Center site the pressure was acceptable at the high point (25 PSI) when we performed the final inspection, but the following year the pressure dropped. We measured the static pressure at the upstream side of the backflow preventer. Operating an average sized zone at 50 gpm the pressure dropped 20 PSI, which was three times the loss we had calculated in the service.

The contractor dug up 25 feet of 2-inch "L" copper service line to check for kinks and disconnected the service to check for obstructions within the pipe. Every thing was fine, so we suggested to the government that a valve was partially closed, or there was an obstruction in the 4-inch water main that provided water to this site.

After reviewing our hydraulic calculations and test results they agreed. The other site was a medical facility where the static pressure was 74 PSI. After the system was installed the operating pressure was too low. After checking the hydraulic calculations on several occasions, along with the static pressure no problems were detected. The system was equipped with a flow sensor and a water meter downstream of the backflow preventer, allowing us to monitor flow rates while checking pressure at various points in the system.

We also performed operating and static pressure tests at the upstream and downstream side of the backflow preventer. These tests indicated an average static pressure of 74 PSI and an operating pressure at the same point of 64 to 67 PSI. The losses were 7 to 10 PSI at 30 to 45 gpm, instead of the 0 to 1 PSI loss that was calculated during design. The Water Department did a flow test, found a valve partially closed and corrected the problem.

Pressure testing is a great diagnostic tool for troubleshooting the irrigation system. We have

This photo shows the effect of a lateral line leak at 140 PSI. Electrical voltage can have a greater consequence than water pressure and should be treated with respect. Testing electronics in an irrigation system takes expertise, patience and often a little luck.

identified faulty backflow preventers, pressure reducing valves and source problems using these methods. Remember, irrigation system pressure is like blood pressure, both need to be checked periodically — more often with age.

ELECTRONICS. To test the electronics in the irrigation system use a volt-ohmmeter capable of reading 120-volt AC (alternating current), 24 volt AC and 0 to 1 megaohm resistance.

Start by testing the power supply for the irrigation controller. Using a voltmeter, the reading should be between 105 and 125 volts. If no power is measured, check the circuit breaker or fuse. If the power source is good, check the circuit breaker or fuse in the controller and reset or replace as required.

Next check the controller's 24-volt output at the wire connections. Turn on station number one and use the voltmeter to test the voltage between station one and the common connection or wire. The voltage is usually in the 24- to 30-volt range because most system valves require 24 volts.

If 24-volt power is not present, the problem is most likely located in the controller. Check the backup battery for power and verify the output voltage of the 110V/24V transformer. The range should be 24 to 48 volts at the output or secondary side of the transformer. If power is above or below this range, turn the power off and measure the resistance across the primary or 110-volt side of the transformer.

The transformer is bad if the ohmmeter indicates 0 ohms resistance. If the station terminals show no output voltage — and unless you're an electronic genius — send it to the distributor or factory authorized service center.

If the proper voltage is measured, the problem is either the solenoid on the control valve or the wiring between the controller and the valve.

If all control valves quit at the same time it could be a broken common wire or a cut in all of the hot wires. When wire insulation is cut or damaged and the wire makes contact with another, the wire is shorted. If the hot wire for one valve is shorted to another, both valves may operate at the same time.

When a hot and common wire are shorted, the breaker or fuse will blow. In addition,

A broken head lowers the pressure in this zone, but not low enough to operate at 30 PSI. Adjusting the pressure reducing valve is one way to correct the high pressure. Checking the system after each mowing helps identify any new problems with the system.

damaged wires found in wet soil that are not in direct contact will ground and cause the same problem as shorted wires.

The solenoid and field wiring can be tested from the controller using an ohmmeter. Disconnect the common wire from the controller and attach to one lead from the ohmmeter. Touch the other lead to each station terminal and note the readings.

Most good solenoids will measure between 20 and 60 ohms, a shorted solenoid will read 0 to 5 ohms and an open solenoid will read higher than 60 ohms — usually at infinity. Replace the solenoid if it reads shorted or open. Two valves that are wired to operate simultaneously will have approximately half the resistance of a single solenoid or between 10 and 30 ohms.

If the solenoid is good, check the power at the valve with a voltmeter. The reading should be 24 volts or higher. If voltage is good, look for a mechanical problem in the control valve. If voltage is low or non-existent, then it's probably a wiring problem between the controller and the valve.

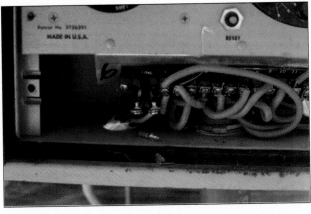

The black and white wires in the middle of the photo are the 110-volt power supply for the system. The exposed copper power supply wires are anchored to broken terminal strips — an accident waiting to happen.

To test for shorted wiring, disconnect the hot and common wires from the controller and the solenoid. Use the ohmmeter to measure the resistance between the hot and common wires. If resistance is low, a short is present. To detect a ground fault, test the resistance between each wire and an earth ground. If the reading is under 100 ohms, a ground fault is likely.

Check for broken wires and bad connectors by disconnecting both wires at the control valve and by pushing bare ends into the soil to maintain an earth ground. At the controller, disconnect both wires and test the voltage from the hot and common wires to an earth ground. The ohmmeter will show a low resistance if the wires are good.

The easiest way to locate a shorted wire or connector is with a ground fault locator. (Most irrigation equipment distributors and dealers sell and rent these devices.)

PUMP MAINTENANCE. When the pump quits not only does the landscape turn brown, but clients scream and you generally lose the contract. Proper pump maintenance will ensure fewer headaches, lower operating costs, higher efficiency and a longer lifespan.

The pump should be inspected at regular intervals depending on total weekly operating hours. Pumps are a major investment that can result in higher energy costs and lower water pressure if not properly maintained. Efficiency losses due to lack of lubricant and wear will add up to increased costs and a shorter pump life.

The tee attached to the pressure gauge has been inserted between the nozzle and the pop-up stem. This device indicates an operating pressure of 40 PSI and, compared with design pressures, indicates whether the system is operating properly.

A regular maintenance check should include an inspection of the pump and the surrounding room or enclosure. Check for water and oil leaks on the floor, then turn on the pump and check again. After operating for 30 minutes or more, feel the pump and motor for operating temperature and vibrations. Check for oil and water leaks. Is the electrical odor normal? Is the pump running hotter than usual?

If the pump motor has Zerk lubrication fittings, it will require regular lubrication. Motors that are 5-hp and smaller generally have sealed bearings that don't require lubrication. Unless otherwise recommended by a motor or pump manufacturer, the proper lubricant is EP-2.

To lubricate electric motor bearings turn off the pump, remove the drain grease plug and with a grease gun, pump the lubricant into the fitting until it comes out the drain plug. Leave the drain plug open for a few days to allow the excess lubricant to drain out.

The bearings will run unusually hot during the first 15 to 20 minutes of operation after lubrication and later return to normal operating temperature.

Pumps that are frame-mounted (separate motor) are normally greased through the bearing cover with the excess draining out the bottom or through a drain plug at the bottom of the frame. Check and clean the bug screens installed on the motor vents.

Make sure the pump housing is well ventilated. A hot room will reduce the motor life by 50 percent for every 18 degrees in temperature above the motor nameplate rating.

CHECK FOR VIBRATIONS. A vibration could be the start of a bearing failure or a misaligned drive coupling or connecting pipe, while a noise in the pump might indicate a bearing failure. If the bearing is going bad, replace it immediately to avoid additional damage.

If a diagnosis cannot be made call in experts for help instead of waiting for additional damage. Have an electrician check the pump control panel periodically for worn contacts and loose connections.

QUESTIONS:

1. What is an easy way to check the operating pressure in an irrigation system?

_____

2. What causes higher operating pressure in the system?

_____

3. Can the solenoid and field wiring be tested from the controller?

_____

4. What is the easiest way to locate a shorted wire or connector?

_____

5. How does temperature affect pump operation?

_____

6. What can cause the pump to vibrate?

_____

7. How often should system operating pressure be checked?

_____

8. What can cause changes in water pressure?

_____

9. If all control valves won't operate, and the controller electrical output is correct, what is the problem?

_____

10. How is a ground fault detected?

_____

(Answers to these questions are found on page 256.)

18

SENSIBLE SENSORS

RAIN SHUTOFF DEVICES are the most common sensors in use today. Rain sensors that interrupt irrigation have been available for many years, but a number have failed because debris — such as leaves or roofing materials — easily enter the sensing area, holding moisture in. This results in false interruptions and dry lawns.

Some newer devices work well, i.e., they are maintenance free (without a cup to collect debris) and adjust from 1/8 inch to 1 inch of rainfall. Sensors should not be installed where the client's irrigation system or the neighbor's irrigation system can fill the catch device. If installed properly, rain sensors can eliminate time consuming trips to the landscape to determine if the irrigation system is applying the right amount of water.

Additionally, freeze sensors prevent irrigation during freezing or near freezing temperatures and reset automatically when the temperature rises above 39 degrees Fahrenheit. This prevents the chance of a liability claim from ice remaining on streets and walks in colder climates.

FLOW SENSORS. Flow sensors are used in irrigation systems to calculate water application, to identify excess flows or leaks and to log total water consumption. Flow sensors can automatically turn the system master valve off in the event of a mainline leak. Additionally, if a lateral line break occurs or a nozzle is missing the zone will skip automatically. This results in tremendous water savings.

Flow sensor accuracy requires a minimal amount of turbulence in the vicinity of the propeller or paddle wheel. A change in the direction of general flow or the flow through valves usually increases the amount of turbulence found in that part of the system. To avoid this, always install a straight length of pipe (the same size as a sensor input and output) that is 10 pipe-size diameters long on the intake side of the sensor, and another length of pipe on the outlet side that is six pipe-size diameters long.

If the paddle wheel is installed in a 2-inch body or tee the upstream length of pipe should be 20 inches and the downstream length 12 inches. This reduces turbulence and improves

accuracy. Some flow sensors, depending on how they are used, also function as a master control valve and/or pressure control valve.

WEATHER STATIONS & ET GAUGES. Weather stations can monitor climatic conditions such as wind speed and direction, solar radiation, air temperature, rainfall and relative humidity. This information is then used to determine daily evapotranspiration rates, forecast the next irrigation and the amount of water to be applied.

A good weather station can cost from $10,000 to $15,000 but save up to 20 percent on water costs through efficient scheduling. Weather stations are not a panacea, but can aid the operator in scheduling irrigation. Weather stations read ET conditions at a specific location that may not represent the same conditions (i.e., rainfall, cloud cover) at another area of the site.

Synthetic ET gauges are modified atmometers that simulate the ET of turf and other crops. These gauges are tubes filled with distilled water and topped with a green covering placed over the top allowing water vapor to move up at same rate that turfgrass leaves transpire.

ET gauges range from $200 to $1,200 (manual read to electronic) and can be almost as accurate as weather stations for a lower cost. These gauges will save about the same amount of water as the weather station. However, the gauges must be filled with distilled water every month or two, and brought inside during the colder months to protect from freezing.

A flow sensor could have detected these missing nozzles, shutoff the water and advanced to the next zone, as well as alerted maintenance personnel of a leak on a specific zone.

SOIL MOISTURE SENSORS. Why use soil moisture sensors? We believe the most efficient way to control the amount of water applied to the landscape is to measure the plant water used within the root zone.

It is time consuming for maintenance personnel to check the landscape site several times a week or even daily to determine if the irrigation system is applying the right amount of water. Rainfall, cloudy days, high temperatures, wind, cold spells and humidity can affect the amount of water required by plants. Soil moisture sensors, along with solid-state control systems, can automate this function and provide the right amount of water for the individual plant and/or property.

There are many different types of soil moisture sensors on the market today. The most commonly used sen-

Soil moisture sensors could have prevented overwatering at this site that resulted in water coming up between the concrete slabs on the road.

sors in landscaping are tensiometers, solid-state tensiometers, gypsum blocks and electrical resistance blocks.

Tensiometers measure the matrix potential or capillary tension in the soil. This is similar to the force a root must exert to take in water from the soil. Gypsum and electrical resistance blocks also measure the matrix potential and consist of electrodes that are embedded in soil or gypsum.

Residential sites can be controlled with one or two sensors, while commercial and golf course sites may require up to one sensor for every two to four zone control valves. These types of sensors should be adjustable from the irrigation system controller and should meet the following criteria:

- Sensors should be adjustable from the controller location.
- The system should provide manual and programmable sensor override.
- The sensor should be maintenance free.
- Equipment should withstand freezing soils, if required.
- Sensors should be corrosion resistant.

Before making an investment in your own soil moisture sensor system, check with others who have been working with sensors for at least two years to see how well they have performed in the landscape.

For the last 10 years we have experimented and collected data with a number of different types of soil moisture sensors. We have installed them in my flower garden and watched and wondered about their effectiveness. Many failed to work properly.

In the last couple of years, however, the quality and reliability of soil moisture sensors has improved significantly. The industry has learned a lot from earlier mistakes and with new technology has been able to design better components. Today, soil moisture sensors can be used with several central/satellite control systems. Some sensors can even be installed in existing irrigation systems with little additional wiring.

In 1989 we designed an irrigation system for an 11-acre park — The Trails at Westlake in Broomfield, Colo. The park was built by a developer and then turned over to the city upon completion. The City of Broomfield was concerned about operating costs — the cost of water in particular — which was $1.63 per thousand gallons or $1.22 per hundred cubic feet (1989 figures) for an annual usage of 6,571,000 gallons with a total annual cost of $10,710. We suggested a soil moisture sensing system to help conserve water and asked if the city would help pay for it.

After explaining the system to the city, they approved the soil moisture

ACCURATE WATER APPLICATION

WHY BE concerned about applying only the amount of water that the plant requires for healthy growth? What are the benefits of this approach?

- Conserving more water for the future.
- Saving money on lower watering costs.
- Reducing fertilizers leaching past the root zone.
- Reducing runoff and pollution.
- Minimizing damage to pavement, walks and buildings.
- Reducing customer complaints of over watering.
- Decreasing the number of drainage problems.
- Producing healthier plants with less disease.
- Producing a beautiful landscape.
- Lowering maintenance expenses.

sensing system and agreed to pay for the system by reducing the cost the developer would have to pay to tap into the city water system. The tap fee or license was $198,601 for a 4-inch water tap irrigating a little more than an 11-acre site.

Including installation, the additional cost of the soil moisture sensing system and controls was $22,297. The city credited this amount against the tap fee, along with another $31,269 for future anticipated water savings for a total of $53,566.

The sensors, along with the rest of the irrigation system, were installed in the fall of 1989. Most of the fine-tuning required to bring the system on line wasn't completed until early summer the following year. Tim Mason, the parks superintendent for the city of Broomfield set the controllers for this park at basically the same timing as the rest of his park sites. In the late summer and fall he reported a 35 percent savings in his water use. The city will see a payback in the form of reduced water costs in less than five years.

We did not expect the savings to be this high, since Mason had already done an excellent job of controlling and conserving the water used in the entire Broomfield park system. This is one of many examples of how responsible system management can result in tremendous benefits.

In a different city, another system was installed in street medians in 1989. It was unique in that it was the first central/satellite control system in the area using soil moisture sensors for system control. With its shrinking budget, the city could not afford to check controllers every week (eight controllers and one sensor for every three to four control valves), but the central control system allowed for easier operation and monitoring.

The moisture sensors for the medians could be adjusted via phone modem from the city offices. Both the central program and the satellite controller maintained a historical log of irrigation events such as the time of occurrence, start and finish times, the amount of time programmed vs. actual watering time, moisture set point, moisture reading and zone flow rates. The sensor could be overridden if fertilizer was applied or maintenance needed to be accomplished. The sensor system could be programmed for the override mode for certain hours of the day or week. Afterward, irrigation events could be printed for comparison purposes and permanent records.

Flow sensors and master control valves were installed at each backflow preventer which allowed the central/satellite control system to automatically shut the system off in the event of a mainline leak. It also detected excess flow (broken riser or head) in a zone and routinely skipped to the next zone leaving a warning message at the central computer. This, in turn, could activate a paging device.

How do you evaluate the need and justify the cost of soil moisture sensors for your site? First, a word of caution. If the system was poorly designed and/or improperly installed or maintained, sensors may be of little value until the system's efficiency is improved.

While your need may be to reduce high water costs or to provide for a healthier landscape, you must first determine the payback period. Ascertain the cost of the moisture sensor system and amortize it over five to 10 years. Compare these figures to potential savings in water costs, labor, reduced asphalt repair, etc.

Sensors will probably make your life easier in many ways. Primarily, sensors will save you money, water and provide a healthier landscape with fewer maintenance requirements.

QUESTIONS:

1. Why have many rain shutoff devices failed?

_____

2. How can flow sensors help conserve water?

_____

3. What is critical for accurate sensor readings?

_____

4. What are ET gauges?

_____

5. What are the most common types of soil moisture sensors?

_____

6. How do soil moisture sensors work?

_____

7. Why use freeze sensors?

_____

8. What needs to be done upstream and downstream of the flow sensors?

_____

(Answers to these questions are found on page 257.)

SLOPPY DESIGN CAN COST YOU A BUNDLE

TAKING SHORTCUTS IN irrigation design and installation may save you money now, but result in legal fees and costly settlements in the future. Property owners are sometimes left with no recourse but to sue developers, designers, contractors and even manufacturers for inefficient and poorly designed irrigation systems.

Many irrigation designers stretch head spacing, ignore good hydraulic design, neglect water pressure controls and specify residential equipment in commercial applications to name a few glaring problems. This is often done in the interest of cutting costs and increasing profit. Whatever happened to ethics and quality?

In light of sloppy design, property owners are becoming more concerned about the cost of maintaining the landscape. Water costs are increasing as are the penalties enforced to prevent waste. Water costs in the Washington, D.C., area, for instance, average more than $3 per thousand gallons. Likewise, the Denver area reports rates as high as $5 per thousand gallons. One large housing development in California paid out more than $300,000 in penalties for excess water use in 1993.

Maintenance costs are high and will continue to increase because most of the irrigation systems in the United States have been designed for the gains of the contractor or the developer, and not the economic needs of the individual water user or the plants.

Because environmental concerns and the water conservation ethic are here to stay, the irrigation industry needs to improve its reputation and take quality and water conservation more seriously.

THE CONSEQUENCES. The Crossings homeowners' association in Denver (140 condominium units), and the steps taken to resolve design flaws is an expensive example of what can happen following a poor irrigation design.

In 1991, an evaluation of The Crossing's irrigation system was initiated to determine if the turf was being over watered, and whether the design was the cause of drainage problems and structural damage. The site was mostly level; covering about five acres of turf area.

An investigation revealed numerous problems with the design and original installation of the irrigation system. The system did not provide uniform coverage, resulting in over watering or "flood" irrigation in some areas in an effort to keep the grass green in other sections

Our findings indicated that the head spacing was stretched in many cases. Most of the pop-up spray heads were spaced at 18 feet to 23 feet (65 percent to 80 percent spacing) instead of the manufacturers' recommended 15 feet. Rotor heads were spaced at 44 feet to 48 feet in lieu of the recommended spacing of 38 feet. There was a great deal of over spray onto the buildings, asphalt, roads and tennis courts.

A rotor and pop-up spray head on the same zone is a mistake. The spray head applies three to four times as much water as the rotor head, resulting in poor uniformity, overwatering and waste.

Many of the pop-up spray head zones had high operating water pressures ranging from 40 PSI to 50 PSI instead of the required 30 PSI. Some of the heads were operating at pressures below 30 PSI — in the 15 PSI to 20 PSI range. These high and low water pressures resulted in a reduced radius of coverage from 15 feet at 30 PSI down to an 11-foot radius at 15 PSI.

The impact rotor head pressure ranged from 20 PSI to 60 PSI resulting in both high and low pressures. These defects caused a distortion of the spray pattern, reduced radius coverage, poor stream breakup and variable rotation speeds, all of which resulted in poor uniformity of coverage and wasted water.

The irrigation system was supplied by three 3/4-inch water taps, with 40 feet of 3/4-inch "K" copper service pipe and 3/4-inch water meters. The system runtime was calculated based on 1 1/2 inches of water applied per week during July. The system runtime measures out at 23.33 hours per day. Consequently, The Crossings received enormous water bills that reflect daytime watering, and residents were deprived of the use of turf areas for summertime activities.

A check with the landscape plans while designing the system would have indicated that rotor heads were the wrong choice for this area.

Many of the irrigation zones had a flow rate of 50 to 59 gallons per minute (gpm). At 59 gallons per minute (19.6 gpm per meter) the velocity of the flow averages 14.43 feet per second (fps) in the 3/4-inch "K" copper service line. Thus, one meter — because of its location — will always operate at a velocity of flow greater than 15 fps, which is excessively high and increases the probability of failure.

The Crossings' flow rates ran up to 11 fps through the 3/4-inch backflow preventer. This exceeds industry standards and the manufacturers' recommended velocity flow rate of 7.5 fps maximum. The purpose of establishing a recommended flow rate is to prevent water hammer within the system which causes a rapid deterioration of the system (pipes, valves and fittings) resulting in leaks.

Irrigation water consumption data based on meter readings at the project for 1988, 1989 and 1990 indicate the following as the total annual irrigation usage:

	gallons of water	inches of water
1988	6,278,000	45.27"
1989	6,554,000	47.26"
1990	8,522,000	61.46"

Turf water requirements for this area do not exceed 30 inches per square foot, or 4,160,000 gallons on an annual basis. Usage above 30 inches per year is due to the deficiencies in the design and installation of the irrigation system.

Proper and efficient irrigation design for this site reduces irrigation water consumption by as much as 42 percent, or 2,958,000 gallons (three-year average). This results in a water cost savings of $3,875 per year at 1991 water rates.

Because of the deficiencies in irrigation design and installation, the homeowners' association at The Crossings successfully sued the developer for damages and received a large settlement.

Poorly designed irrigation systems, such as at The Crossings, can cause severe damage to structures, asphalt paving and plants as well as waste water. Americans are spending millions of dollars every year to repair and replace the paint, siding, asphalt, concrete and plants that are damaged from overwatering and overspraying.

A leak cost this owner $64.80 (21,600 gallons) each day for about a month, and totaled almost $2,000 dollars (650,000 gallons).

Asphalt and paint will deteriorate when water is frequently applied. The combination of swelling soils and settling have damaged all types of structures, resulting in huge insurance settlements and higher rates. Water damage from spray through open windows, flooding and water seeping into basements has likewise wreaked havoc on homes and the companies that insure them.

Personal injury can also occur from heads and valve boxes installed too low in turf and athletic fields. We have seen several cases where people have stepped into vaults and valve boxes, and sustained serious injury because of unsafe lids. People have also tripped and fallen on sprinkler heads and risers. Water and ice on walks and roads have caused numerous vehicular and pedestrian accidents.

High pressure combined with lateral line surge blew this head right off the threaded fitting. Pressure control and check valves under the heads could have prevented this waste.

PRESSURE CONTROL. Many designers do not understand the importance of pressure control. High pressure causes pop-up spray heads to mist and shriek from the high velocity of flow, and leak around the heads and wiper seals. We have seen rotor heads operating like "machine guns" and spinning tops, creating a bank of mist that drifts away from the area for which it was intended. This dramatically reduces the longevity of the head in the irrigation system.

Low operating pressure causes the water to explode into larger droplets producing soil compaction and reducing the effective radius of coverage. High pressure causes the water to explode out of the nozzle into a higher number of tiny droplets that range in size from 1.0 mm (moderate rain) to 0.10 mm (mist).

A 1.00 mm drop falling from 10 feet in a 3 mph wind will drift 5 feet, while a 0.10 mm drop will drift 50 feet. This reduces the effective radius of coverage and causes the water to appear as a drifting mist. This mist will evaporate much faster than larger droplets, and will easily drift away from the irrigated area. Imagine the water wasted by drift in a 5 or 10 mile per hour wind.

The ideal operating pressure (and therefore adequate droplet size to meet the needs of plant materials) for small pop-up spray heads is 25 to 30 PSI. A pressure of 30 to 50 PSI is recommended for most rotor head applications.

PRESSURE CONTROL REMEDIES. Water pressure can be easily regulated. To correct high pressure use pressure reducing valves and pressure loss in the pipe to provide optimum pressure at each head. Use pop-up spray heads with a pressure reducing device (set for 30 PSI) installed as an integral part of the unit.

Pressure reducing valves can be installed at the electric control valve to control pressure for the rotor head zones. Plastic preset PRVs can be installed under each rotor head. When the static pressure in residential and commercial systems is more than 75 PSI, we prefer to install an adjustable PRV at the point-of-connection to protect the system from unnecessary water surges.

High water pressure can also cause surges in lateral lines, especially if the lateral is drained or partially drained after every cycle. This results in damage to the equipment, water leaks and a reduced system life span. Install heads with check valves to prevent low head drainage and to save water.

Low water pressure can be avoided by carefully calculating hydraulics for every design. If possible, allow for a 10 to 15 PSI supply pressure drop as a safety factor.

Irrigation system uniformity is the bottom line. Several town house associations in the Denver area received large cash settlements from designers and contractors for the improper design and installation of their irrigation systems. The contractors at fault all did the same thing: they stretched the head spacing to 70 percent and 80 percent of the diameter of coverage when most manufacturers recommend spacing at 50 percent of the diameter.

The result is unacceptable uniformity causing damaged turf and asphalt from the over watering required to prevent the turf from wilting and dying. Proper spacing for any head should be no greater than the manufacturers' recommendations, plus necessary adjustments for wind conditions. Uniformity is important in turfgrass where every square inch has roots and a shallow root zone in which 75 percent of the roots are often found in the top 1 to 2 inches of soil.

The essential components of a quality irrigation system include equipment that is high in quality and performance, low in maintenance, long lasting and water conserving. Selecting the right equipment for an irrigation project is vital to its long-term success. Many contractors and designers select equipment based on price or friendship instead of considering the quality, performance and maintainability of the equipment.

For example, head selection for turf areas should require a minimum pop-up height of 4 inches. If the mowing height is more then 3 inches, use a 6- or 12-inch pop-up height. This length is necessary because of generally higher mowing heights in turfgrasses and normally occurring turf buildup. For appropriate equipment uses it's recommended that slotted brass nozzles for small spray heads are used. Plastic nozzles usually apply little water within 12 inches to 18 inches of the head. Brass nozzles have a slot cut into the nozzle below the main outlet that provides water for the area immediately around the head.

Smaller, plastic impact rotor heads expose the system to more damage, while the use of stronger materials or smaller diameter heads minimizes damage from equipment, vandalism and injuries.

Swing joints are necessary to protect sprinkler heads from damage. Flexible, kink resistant tubing or pre-manufactured PVC swing joints are the best choice. Avoid PVC swing joints you assemble yourself; pre-manufactured swing joints prevent leakage.

CONTROLLERS AND VALVES. The single most important selection criteria for an automatic controller is ease of operation and simplicity. If you need an instruction booklet in order to operate it, look for another controller. Other important features contributing to irrigation efficiency are accurate timing from hours to seconds, multiple repeat cycles, flexible day scheduling, sensor input and water budgeting.

To apply the correct amount of water, the timing should be accurate within seconds. Multiple repeat cycles reduce runoff and improve infiltration rates. Flexible day scheduling is useful for mowing, special events and so on particularly if water restrictions are mandated. Insist on a memory retention in the computer chip instead of replacing a rechargeable backup battery every year.

Look for valves that have an internal manual bleed so the valve box won't fill with water. The small ports in the valve should be clog resistant and/or self-cleaning when dirty water is present. An encapsulated solenoid with a captured plunger helps the life span and reduces maintenance headaches.

Select valve boxes that withstand compaction and weight against the side and top of the valve box. Valve boxes have been known to cave in on the sides (12-inch rectangular) because of the lack of structural strength.

The recommendations in this book are only a start. Each designer should determine the requirements for a specific project based on quality, efficiency and longevity. Products that work well can save maintenance dollars, improve the reputation of our industry, reduce headaches, conserve water and provide us with a beautiful, healthy environment.

Quality control is vitally important throughout the design and installation of an irrigation system. Contractors often take shortcuts and substitute quality for economy or profit.

Every designer should insist on reviewing the installation process, answering questions and evaluating the final installation to make sure the system was installed correctly and operates properly. Contractors with designers on staff should also follow this practice as a simple quality control procedure. Independent designers should include site observation services for all their designs both to protect the client and themselves and to maintain a better reputation.

We can reduce our liability considerably if we improve the quality of design in our industry. How do we do it? More and better education. Design systems with more care and concern. Design for higher efficiency, a longer life span and lower maintenance costs. People demand, expect and deserve it.

QUESTIONS:

1. What is the maximum allowable velocity for "K" copper service lines?

_____

2. What causes large water droplets, soil compaction and reduces the effective radius of coverage?

_____

3. What device is the best to use when controlling pressure?

_____

4. What are the primary concerns of owners and managers?

_____

5. What causes water to mist and reduces the effective radius of coverage?

_____

6. What can cause severe damage to structures and asphalt paving?

_____

(Answers to these questions are found on page 257.)

20

PRECISE BIDDING = PROFITABILITY

DON'T LET THE unknown ruin what you think is a perfectly good bid. Prepare a thorough bid by first visiting the job site and uncovering potentially hidden problems.

Six years ago, an irrigation contractor in New York called me to discuss his business. It seems his firm was losing money on most of its projects and he wanted me to review his operation in an effort to help him correct the problems. Three weeks later I was in New York observing the firm's installation methods, estimating processes, accounting procedures and general methods of operation.

The firm's installation techniques were adequate, but morale was low stemming mainly from the installer's frustration with the profitability of the irrigation division — or lack thereof. The crews revealed they were encountering rock or difficult soil conditions on many of their projects resulting in slower installation times and high trencher maintenance costs. On top of that the crews had no performance criteria or labor budgets. Likewise, the owner had little information to use in evaluating employee performance.

A look at estimating and cost accounting procedures revealed the real culprits. Labor and material job costing were non-existent. This left the estimator guessing about labor costs, and gave him no sound basis for estimating future work. Apparently, the estimator was extremely busy leaving him little time to look at the job site prior to bidding the job. Visiting the job sites and digging a few holes would have disclosed the soil type. With proper job costing he could have then bid the project accordingly and for profit.

The contractor changed some of his procedures and, within a year, the irrigation division was showing a profit. This story is typical of how many irrigation contractors fail to manage their estimating and cost accounting methods.

AVOIDING PITFALLS. What are the components of a good, solid estimate, and how can some of the pitfalls contractors experience be avoided? Many questions should be asked before bidding a job, but one should always come first, "Does this project fit my business and can it be profitable?" Contractors need to learn to be selective about the type of projects they bid and avoid going after every potential project.

Additional considerations, "Will I be a prime contractor or a subcontractor?" I would rather be the prime contractor because of faster payment, improved communications and greater control of project management issues. "Is the owner financially capable of completing the project? "Is the designer fair to deal with?" "Is a retention held on the monthly payments and will payments be prompt?

Deciding between the roles of prime contractor or subcontractor is a control issue, and grabbing the prime contractor position gives the lead firm

An accurate estimate allows a contractor time to install the system properly for good performance and longevity.

more control over scheduling, payments, negotiations and disputes. If payments take 45 to 60 days and retention is 10 percent until acceptance, interest charges should be added to the bid.

Look at the size and type of proposed irrigation system to see if your firm has the experience and financial capability to do the project efficiently and profitably. Check the bonding and insurance requirements as well as taxes, licenses, permits and additional costs that may come up during the project.

After making the decision to put an estimate together, perform a job site inspection. Check the drive time from your office to the site to see if it is longer than an average drive time. If so, add that time to your estimate. While you're at the site dig some holes to check the soil conditions. Don't bypass this step to save time. It reminds me of the time we bid a golf course in Tulsa, Okla. After looking at the site, we sensed there might be some rock problems. So we rented a backhoe and dug several test holes and discovered that the entire site had a limestone shelf 12 to 24 inches under the soil surface. Such an oversight could easily bankrupt a contractor.

Once the soil and ease of trenching/excavating is determined look for steep slopes or poor drainage areas that may require hand digging or result in slower installation. Survey the area surrounding the site for vandalism or potential vandalism. Nearby schools can result in children traversing the site leading to future problems.

Other thoughts to help refine your bid include: Does the site have a lot of pedestrian or vehicular traffic? This may require fences and barricades which cost money. Will other construction trades be involved in the project, resulting in delays and several move-ins?

This can cause scheduling nightmares and added expense to the job. Are all areas accessible to the equipment and is power available at the automatic controller? Who installs the sleeves and how deep will they be? Sometimes locating sleeves takes more time than installation. Answering all of these questions prior to bidding will help price your job profitably.

This site inspection revealed a gravelly, sandy soil and an existing 3/4-inch water meter and backflow preventer. The existing soil has little water holding capacity and will be difficult to trench.

PREPARING THE BID. Set up an estimate spreadsheet with the following headings across the top: quantity, size or measure, description, material unit price, material total, labor unit price, labor total, total materials and labor. Next, perform a quantity material takeoff. As you count heads, valves, etc., use a colored pencil to mark each one so you know which ones you've counted. This is helpful if you're interrupted. Color all the lateral lines using a different color for each pipe size, making judgments faster and more accurate.

Begin with the system quantities in the same order they will be installed. This practice helps eliminate omissions and gives you a better job understanding. Likewise, when listing the items on the bid form, do so in the order they will be installed to prevent omissions.

Beginning with mobilization, follow through with the point-of-connection, backflow preventer, specialty valves, sensors, trenching and backfill, mainline by size, quick coupler valves, manual drain valves if necessary, automatic control valves, automatic controller, wire, lateral lines by size, heads by type, turnover material (keys, spare parts, etc.), testing and adjusting the system, cleanup, as-built drawings and winterization, if required.

Once all the bid items are listed, contact your local supplier for current equipment prices. Enter the unit prices on the irrigation estimate sheet. Job costing records will provide the amount of labor required to trench and install the equipment. For example, it might take 15 minutes to install a pop-up head or 1.5 minutes per foot to trench, backfill and compact the trench.

If the average hourly labor cost is $12 per hour, the labor unit price per head is $3 for the pop-up spray head, and 30 cents per lineal foot for the trenching. The average labor cost should include payroll taxes such as FICA and state and federal unemployment taxes. Workmen's compensation insurance and the company paid portion of medical insurance should be included as well as retirement benefits, if applicable.

Many contractors forget about overtime expenses for a seasonal business. If your crews work 50 hours a week, overtime premiums need to be added to the average labor cost. Holiday, vacation and sick pay should also be included, if applicable.

Pitfalls to a beautiful landscape can be avoided by solid estimating and design methods. Uncovering all of a site's potential problems before a project begins leads to a more profitable job.

After labor totals are entered in the estimate under labor quantities, extend all of the material and labor units, enter totals both down and across the spreadsheet and add the total columns both down and across. This is important. Many bid errors are found using this double-check method.

Next, add up the material and labor totals from all of the irrigation estimate sheets and enter those totals on the irrigation bid summary sheet. Compute the sales tax on material (if applicable) and round up to the nearest dollar. Do this with all your extensions and calculations to speed up the process.

Now determine your overhead cost. Overhead should include all business expenses not listed on the irrigation estimate sheet or in the labor cost and labor burden mentioned earlier. If you're not sure what to include or how to go about calculating overhead, ask your accountant for help. If the estimate is not accurate you could lose a lot of money or obtain few jobs.

For example, if your overhead is 25 percent of total sales, that percentage must be converted to an add-on percentage which, in this case, equals 33 percent. Calculate the overhead on the labor and material total and enter the total.

Profit should be a minimum of 10 percent of your sales price or an add-on of 11 percent. Because of the high risk involved with irrigation systems and all of the unknowns involved with weather, underground utilities and excavation, a minimum profit goal of 15 percent is recommended.

Enter any miscellaneous costs i.e., list subcontractors at cost and add a mark-up of 10 percent to 20 percent to cover the cost of dealing with the subcontractor. Add a mark-up on subcontracts just in case they go out of business and your firm is stuck paying more to complete the work. If a performance bond is required add your cost for a bond and total the estimate. Give the estimate to someone else to check the math and review the estimate.

REFLECTION. Ask yourself if you feel the estimate is accurate. Do you need the work, can you handle the project without a lot of problems and will it be easy to do? If you have any doubts about the project or if the market will allow a higher price add more money. At one time, we prepared an estimate totaling $25,000 for an irrigation project and added $4,500 to the total because things didn't feel right. The result — we were the low bidder at $29,500. The next highest bidder was $29,900.

Unit prices can be developed for smaller projects to save time. The unit prices cover the point of connection, backflow preventer, pressure reducing valve, mainline, controller, valves and heads. For example, the unit price for a spray head included 15 to 20 feet of 1-inch pipe, fittings, the head and nozzle and the connection under the head. The mainline unit price included wire.

Remember to adjust the unit prices when material or labor costs change.

Accurate estimating is the key to a profitable irrigation contracting business. Implementation of a good estimating and job costing system will keep you ahead of your competition and help your firm be more profitable.

QUESTIONS:

1. Why are site inspections important for the irrigation contractor?

_____

2. What is the best source of determining labor costs for an irrigation estimate?

_____

3. What methods can prevent estimating errors?

_____

4. What factors should be included in labor costs?

_____

5. What should a contractor determine about a project prior to compiling an estimate?

_____

6. Should interest charges be added to the bid?

_____

7. What is the advantage of being a prime contractor vs. subcontractor?

_____

(Answers to these questions are found on page 257.)

21

EVALUATING EXISTING SYSTEMS

GOVERNMENT REGULATIONS, WATER conservation ethics and the continually increasing cost of water have spurred the landscape and irrigation industry to find creative and effective ways to save water. The results are impressive. Improved irrigation design methods and more efficient equipment have created irrigation systems which are easier to maintain and use less water than older systems.

Still, increasing water costs are forcing owners to find ways to reduce water use. The highest water cost we have seen is $5.82 per thousand or "M" gallons ($4.35 per 100 cubic feet). Many cities across the United States have rates that exceed $3 per M gallons. In many areas, the annual cost of water with an efficient system can easily exceed $2,500 per acre — $5,000 to $10,000 per acre if the system is poorly designed or maintained.

WHAT'S AN EVALUATION? A system evaluation is a means of identifying the safety, life span, operating cost and efficiency of an irrigation system and preparing recommendations with cost estimates for renovation or replacement. It includes an audit of the system to determine uniformity and efficiency, as well as recommendations for the best management practices. An evaluation of the irrigation system usually reduces long-term maintenance and operating costs.

What sites generally require an evaluation? Sites that are more than 12 years old should have an evaluation in an attempt to reduce operating expenses. Poorly designed and installed systems may require an evaluation regardless of the age of the system.

QUALIFICATIONS. The Irrigation Association, a non-profit trade organization, provides landscape irrigation auditor training programs and certification testing throughout the United States. These training programs teach landscape and irrigation professionals to perform a catch-can audit to determine the actual performance of irrigation systems and to develop efficient irrigation schedules. This results in better management of irrigation controllers and minimal water use throughout the season. The Certified Landscape Irrigation Auditor (CLIA) designation, for instance, indicates the individual has passed a test and understands the process of operating a system efficiently. However, this doesn't mean a CLIA is necessarily qualified to perform system evaluations, make recommendations for equipment upgrades or design changes.

An irrigation system evaluation exposed this malfunctioning pop-up spray head. Frequent monitoring of the system during operation can pay for itself by reducing damage, saving time and money and reducing water use.

Passing the certification test allows individuals to use the CLIA designation on all correspondence, business cards, etc. This is one of several qualifications required of an individual performing a system evaluation. The evaluator should also have extensive background and training in irrigation system design, installation and maintenance. Completing and passing both the Certified Irrigation Designer and Certified Irrigation Manager certification test offered by the Irrigation Association indicates the individual has the qualifications to perform a system evaluation.

In 1982, Keesen Water Management performed numerous irrigation system evaluations for the Denver Water Department. Sites included parks, golf courses, medians, highways and commercial locales.

High pressure and flowers blocking the spray pattern were corrected using pressure reducing valves and height extenders on some of the 12-inch pop-up spray heads.

The amount of water that can be saved if the systems and scheduling are improved in these and similar locations is amazing. In an area where the average annual evapotranspiration rate is 26 inches, we saw annual water usage ranging from 10 inches in a large park to 160 inches in a boulevard median. Average annual rainfall during the growing season is 10 inches, but more than half of the rainfall is lost because of the sudden downpours that produce runoff.

THE EVALUATION PROCESS. The first step is determining the need for an evaluation by analyzing past water usage and the age of the system. Historical water usage can be obtained from the water purveyor or the owner. A minimum of three years data for averaging is recommend to adjust for annual fluctuations in climatic conditions.

Water usage is measured in either thousands of gallons (M gallons) or per 100 cubic feet (CCF). Most water meters are read — and clients are billed — monthly or bimonthly. Using a spreadsheet, enter water consumption from water bills at the appropriate month and year and total.

If irrigation is on a separate meter and the water is used only for irrigation, enter consumption data using the irrigation meter method. If indoor and irrigation water are on the same meter, use the shared metering method.

IRRIGATION METER METHOD. To determine the amount of inches applied annually use the following formula:

Formula for M gallons:
Annual irrigation water use in M gallons 815 x 1,000 = 815,000 gallons
Net irrigated area in square feet 43,560
Annual water use divided by area covered divided by 0.6234
Note: 0.6234 is the amount of water in gallons in a square foot 1-inch deep.
EXAMPLE: Annual water use in gallons = 815,000
Area in square feet = 43,560
815,000/43,560 = 18.71/0.6234 = 30.01 inches

Formula for CCF:
Annual irrigation water use in CCF 1,089.5 x 100 = 108,950 cubic feet
Net irrigated area in square feet 43,560
Annual water use divided by area covered times 12
Note: 12 is the multiplier used to convert cubic feet to inches
EXAMPLE: Annual water use in cubic feet = 108,950
Area in square feet = 43,560
108,950/43,560 = 2.50 * 12 = 30.01 inches

SHARED METERING METHOD: If indoor and irrigation water are on the same meter, use the following steps to determine the approximate outside water use:

Add up the total water usage for several winter months and divide it by the number of months to determine the average monthly indoor use. Subtract this amount from the cutsomers' water bill amounts for the months in which irrigation occurs (irrigation season).

Example: Bimonthly bills:

December 114 thousand gallons
February   124 thousand gallons

114 + 124 = 238 / 2 = 119 bimonthly average or 59.5 per month for indoor use.

Subtract this amount from irrigation months and proceed to the formula under irrigation meter method above.

IRRIGATION GOALS. The maximum amount of applied water for turf should not exceed 115 percent of the annual plant evapotranspiration rate. If the annual ET is 26 inches, the maximum water use should not exceed 30 inches. Planting beds may require 30 percent to 50 percent less water than turf areas. Large open space turf areas are more efficient and could be at 30 percent to 50 percent of ET or less. In areas where rainfall can be absorbed by the soil, subtract effective rainfall from the annual ET before determining the 115 percent. The ultimate goal is to reduce water consumption to a level below these guidelines.

> # TOOLS TO SPEED THE PROCESS:
>
> Pop-up spray head adapter and PSI gauge
> Pitot tube
> Soil probe
> Remote control
> Volt/Ohm meter
> Small tools

If the maximum annual usage exceeds 115 percent of ET, or 30 inches and does not exceed 150 percent or 40 inches, an audit and a new operating schedule might reduce the use. When usage is more than 150 percent, an evaluation should be performed. These are general guidelines that may vary under certain conditions.

Next, each zone in the system should be operated for at least five minutes to ascertain the physical condition, level of maintenance, operating pressure, efficiency, hazards and functions. This is easily done with the help of a radio remote control unit which allows the operator to turn zones off and on without going back to the controller or locating valve boxes. As each zone is operated walk the area looking for leaks, head malfunctions, potential liability, slow closing valves and pressure problems. Carry a small pad of paper and note the problems by controller and zone (Ex: A-17) so repairs are easily identified by servicemen.

Pressure should be checked to verify the correct operating pressure for each zone and the pressure variation within the zone. Sprinkler head operating pressure should be within 15 percent of the manufac-

The operating pressure is verified using a pitot tube and pressure gauge. If the pressure is too high, take the appropriate steps to control it.

turers' recommended optimum pressure rating for the specific nozzle and spacing used in the design. The pressure variation within a zone should not exceed 15 percent.

Sprinkler heads should be checked for arc and radius of coverage, plugged nozzles, plumb, height above or below turf at rest and in operation, rotation speed, matched precipitation nozzles and malfunctions.

Potential liability issues such as broken or missing valve box covers, settled trenches, low and high heads, overspray on buildings and roadways should be identified and the owner notified by written report.

STEPS IN THE SYSTEM EVALUATION PROCESS:

1. Conduct historical water usage analysis for past 3 years:.
   - operation site inspection of the system
   - catch-can audits to determine efficiency
   - pressure tests
   - leak detection
   - Identify potential system hazards and liabilities
   - check valve closure time
   - equipment condition

2. Review of irrigation schedule
   - determine soil type
   - determine root depth
   - existing schedule

3. Identify potential savings
   - water
   - maintenance

4. Short- and long-term recommendations
   - budget cost estimates
   - operating instructions
   - cost benefits analysis

Other evaluation steps include checking for slow closing and weeping control valves. Valves that are slow closing may operate after the next valve starts its cycle resulting in low pressure for both zones and inadequate coverage. Water that continually seeps from low heads indicates a control valve that is not seating properly. Check for a bad seat or a small particle imbedded in the seat.

Using catch-cans to audit the system performance, the variation in precipitation and uniformity of coverage can be viewed. Once the audit is completed, a more realistic and efficient irrigation schedule can be established.

Don't forget to jot down the existing irrigation schedule and take soil samples. The existing schedule will indicate potential runoff and overwatering. Using a soil probe take at least 10 soil samples per acre. This will help determine root depth and soil type.

After the field work is completed, begin the written report. The report should contain the following information:

- Equipment type and condition
- System operating pressure
- System operating schedule
- Historical water usage
- Short-term recommendations
- Long-term recommendations
- Costs and benefits

The next chapter will continue with the evaluation report and look at recommendations and how they are made. Cost benefits, budget estimates, phasing, what to reuse from the existing system and renovation vs. replacement is also included.

QUESTIONS:

1. What sites usually require an evaluation?

_____

2. What are the necessary qualifications for an individual performing a system evaluation?

_____

3. What is the first step in determining the need for an evaluation?

_____

4. How is water usually measured?

_____

5. What is the formula to convert annual site CCF to inches?

_____

6. Why should slow-closing control valves be repaired?

_____

7. What is the annual irrigation use in inches per square foot for a 60,000 sq. ft. site using 1,850,000 gallons annually?

_____

(Answers to these questions are found on page 258.)

22

APPRAISING IRRIGATION SYSTEMS

OFTEN, IT'S HARD to judge the most effective means of improving system performance. Sometimes replacement of all the heads or a controller will be enough, but in most cases the problems are numerous.

In the previous chapter, discussion focused on the importance of evaluating irrigation systems for renovation or replacement considering the increased cost of water and its limited availability in some locations. The text emphasized the need for proper evaluation of existing irrigation systems. Rather than going to all of your clients and making recommendations for new systems, it's best to perform a system evaluation to ensure the right decision is made based on the viability of the current system, the needs of the client and the long-term impact the irrigation system will have on turf, trees and plants.

Performing irrigation system evaluations takes some practice. It includes identifying the safety, life span, operating cost and efficiency of an irrigation system and preparing recommendations with cost estimates for renovation and replacement.

In a continuing look at system evaluation processes, this chapter includes a discussion on how to write and compile detailed reports.

THE FIRST STEP. Begin an irrigation system evaluation report by describing the equipment by brand name, model, size, age, maintainability, condition and so on. The following descriptions are examples from an evaluation performed by Keesen Water Management:

"Irrigation connections and water supply: Two 2-inch, three 1 1/2-inch, one 1-inch and one 3/4-inch water meters provide clean potable water for the irrigation system. The irrigation points of connection (P.O.C.) to the potable water supply are adequate to supply enough water to complete the irrigation cycles within an 8-hour, 6-day per week time period for the entire site except for Filing 5 which is 11 hours per day, 6 days a week.

Static pressure is high at 80 to 104 pounds per square inch (PSI), except for Filing 1 which is marginal at 58 PSI. No pressure reducing valves were installed. Backflow preventers are one double check valves below grade, one double check valve below grade and one reduced pressure backflow preventer installed above grade. The double check valve backflow preventers do not meet current Denver City Water requirements."

Next, compare the system operating pressures to manufacturers' recommended optimum pressures and explain the importance of proper pressure:

The ideal operating pressure for pop-up spray heads is 25 to 30 PSI. This head is screaming and ready to blow-up. At pressures like this the system won't last very long.

"The ideal operating pressure for pop-up spray heads is 30 PSI at the base of the head. Most of the pop-up spray head zones have a high water pressure ranging from 45 PSI to 85 PSI. A few zones have low water pressure ranging from 7 PSI to 18 PSI.

"This causes a distortion of the spray pattern reduced radius of coverage and losses from wind and evaporation. It also results in poor uniformity of coverage, dry spots, wet spots and wasted water."

A recommended system operating schedule should be included in the report for the owners use after the system is upgraded:

"Numerous soil probes indicate that the soil is a clay soil with a low water infiltration rate (0.10 to 0.20 inches per hour) and a high water holding capacity.

"Because of the soil conditions, every other night or every third night watering is recommended as much as possible, preferably between 10 p.m. and 6 a.m. when there is little wind and generally lower evaporation rates.

"This prevents excessive runoff, allows oxygen to enter the soil, prevents excess leaching of nutrients from the soil, reduces disease and may help the turf root system extend to a deeper level.

"Application rates and system run times are determined by the spacing between the heads and the head flow rate. A good design requires the head spacing to equal the radius of coverage as well as an unimpeded spray trajectory.

"Approximately 80 percent of the pop-up spray heads and 70 percent of the rotor heads do not meet this criteria.

"Assuming improved

The rotor on the top of this berm may have been the only head left in the truck, because the application rates are so mismatched. (The top of the berm requires more water not less.)

operating conditions, the average application rate for rotor heads is 0.50 inch per hour, and for pop-up spray heads 1.5 inches per hour. The water time per cycle for rotor head zones should not exceed 21 minutes, and pop-up spray heads 7 minutes to minimize runoff.

"Peak demand for turfgrass for this site should not exceed 1.5-inches of applied water per week during the hottest portion of the summer. This produces a watering schedule of three mornings (early) a week with 0.5-inch of water applied per day. Three watering cycles are required at 7 minutes each for pop-up spray heads and 20 minutes each for rotor heads.

"This schedule presents a general guideline for peak watering demand after system efficiency is improved by repair or renovation. Individual station run times may need to be adjusted for microclimate characteristics. As seasonal changes occur, the number of watering cycles should be adjusted to conserve additional water.

"The measurement of all water applications for each zone in the irrigation system, using the catch-can test method, will provide for greater accuracy in water/time applications and result in additional water conservation at the site. Aerate the turf areas at least two times per year to allow for better infiltration of water and oxygen into the soil."

A leak below the corner head went undetected for several years. Even after the concrete was patched and the deterioration process started over, the leak went undetected until our evaluation.

SYSTEM EFFICIENCIES. Historical water usage, local turf water requirements, system inefficiencies and potential savings should be explained:

"Irrigation water consumption data based on meter readings for 1990, 1991 and 1992 indicate a total annual irrigation use of approximately 14,689,000, 15,844,000 and 17,660,000 gallons of water or 38.28 inches, 41.29 inches and 46.02 inches per square foot, respectively.

"Annual turf water requirements for this site do not exceed 20.8 inches per square foot. Irrigation system inefficiencies in a properly designed system can produce from 1.2 inches to 6.2 inches of loss which results in 22 inches to 27 inches annually, or up to 9,062,000 gallons for this site.

"Upgrades to the irrigation equipment and design, plus some improved operating techniques will help reduce irrigation water consumption by as much as 51.3 percent (8,598,000 gallons) or $13,241.00 per year at 1993 water rates."

Short-term recommendations are generally repairs for leaks, hazards and minor equipment adjustments. These recommendations should identify the location and the problem, and might appear as follows:

"The following is a listing of site changes or repairs that should be completed as soon as

possible. This listing is by controller and zone.
(Example: H3 refers to controller H, zone #3).

Area I   Filing 1   8552 E. Problem Dr.   2"-meter
Controller "M" (in pump house)
M1 repair rotor west of pump house.
M6 repair two leaks in island at spray heads.

Area II   Filing 2   9051 W. Pretense Ave.   2"-meter
Controller "H" (west side of clubhouse)
H3 repair leak on south side of pool and northwest corner by barbecue grills.
H4 repair non-closing valve and leaks by bicycle stand.
H8 repair three leaks in shrub bed.

General Notes:
1. Trim back shrubs or move heads to avoid interference with spray patterns.
2. Install rain shut-off devices at each controller to better control water application and conserve additional water.
3. The cost for the above repairs and modifications will be approximately $400 to $550

Before making long-term recommendations for the system, first identify which current components are adequate and efficient, then look for potential water and maintenance savings, budget estimates for upgrades and the approximate payback period for improvements.

Irrigation equipment should be evaluated individually and collectively before replacement. This head was not intended to hold up under commercial mowing equipment.

Often, it's hard to judge the most effective means of improving system performance. Sometimes replacement of all the heads or a controller will be enough, but in most cases the problems are numerous.

Keesen Water Management has developed a "System Evaluation Analysis" which is a method of rating seven system components in three or four categories, and then setting guidelines for repair or replacement of any single component. The analysis also indicates when it may be more cost-effective to design a new system.

When you decide whether or not to remodel or install a new system, always ask yourself if it is the most cost effective and efficient method for making improvements.

Long-term recommendations might appear like this:

1. Area I   Filing 1   8553 W. Problem Dr.   2"-meter

"I recommend that the entire system be redesigned and replaced because of problems such as defective pipes, old style heads, leaking wiper seals, the need to raise and/or lower heads, plum heads, head spacing, coverage problems and pressure problems. The controller could be reused if it has enough stations for the redesigned system.

"A new irrigation system would cost approximately $27,000 to $34,000 including design fees and construction observation services. The cost range depends on construction phasing. The system renovation payback period from reduced water and maintenance costs is about 5.4 to 6.8 years."

1. Water costs savings		$2,012
2. Maintenance costs savings		$3,000
	Total	$5,012

The system renovation cost payback period of 5.4 to 6.8 years does not include the cost of borrowing money for improvements.

When the report is completed, set up a meeting with the client to present and discuss the report. Encourage the client to implement the changes and, once improvements are complete, monitor the water usage and maintenance costs on an ongoing basis as an additional service to the client.

The examples explained in this chapter are samples from several different Keesen Water Management evaluations, and should not be viewed as a complete report.

QUESTIONS:

1. What are short-term recommendations?

_____

2. What is the first step in writing the report?

_____

3. How are system operating pressures used in the report?

_____

4. Why is the System Evaluation Analysis used?

_____

5. What are the four factors used in the System Evaluation Analysis?

_____

(Answers to these questions are found on page 258.)

23

GAUGING PUMP PERFORMANCE

THE SINGLE BIGGEST reason why pumping systems fail to operate as designed is because of poor suction conditions between the point where the water enters the pumping system (intake) and the pump impeller. Intake pipe sizes, screens, lifting water from lower elevations, altitude, atmospheric pressure and water temperature are some of the factors that affect pump performance.

Virtually all irrigation pumps use centrifugal force to increase water pressure. The two basic pump types are horizontal end suction centrifugal and vertical turbine. A third is the submersible pump which is a type of vertical turbine with a submersible motor installed below the pump bowls.

The end suction single stage centrifugal pump, referred to simply as a centrifugal pump, is the most common in landscape irrigation because of price and reasonable efficiency. The suction size is usually one pipe size larger than the discharge, and its one-impeller pump discharges water at a right-angle to the water entering the eye of the impeller.

The volute or housing around the impeller can be rotated to provide a variety of discharge directions. The top horizontal direction is the most popular because it allows discharge parallel to the ground surface, and places the priming port at the highest point allowing all the air to be removed from the pump volute when the pump is primed. End suction pumps require priming if they are installed above water level, and if they are not the self-priming type.

Vertical turbine pumps are used in high water pressure sites or when water is pumped from a deep well (submersible pump). Turbine pumps are usually multi-stage because of the increased pressure requirement. Each stage is a bowl assembly with an impeller.

As water leaves the impeller in the first bowl, vanes inside the bowl diffuse and guide the water to the eye of the second impeller. The process is repeated for each additional stage.

Water moves vertically up the inside of a shaft or column and is discharged at a 90-degree angle just below the pump motor. A vertical turbine pump functions as a series of centrifugal

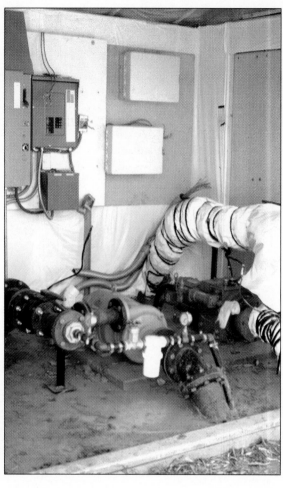

Virtually all irrigation pumps use centrifugal force to increase water pressure. Pictured are two end suction centrifugal pumps.

pumps with each stage boosting the pressure in succession. Vertical turbine pumps do not require priming because the bowls and impellers are installed at a minimum submersion level established by the pump manufacturer.

The difference between submersible and vertical turbine pumps is the installation of the submersible motor below the pump bowls and impellers which are then connected with a short drive shaft. When installing a submersible pump in a lake or canal, make sure water always flows over the motor to keep it cool.

PUMP SUCTION & DISCHARGE CONDITIONS. These conditions, among others, must be calculated accurately to ensure actual flow will meet design requirements.

Calculating system pumping requirements involves using a common formula to determine NPSHA:

$$NPSHA = H_a + (-) H_s - H_f - H_{vp}$$

Where:

H	=	Head in feet.
H_a	=	Atmospheric pressure on the water surface.
H_s	=	Vertical distance from the water surface to the centerline of the impeller measured as static head. Hs is a minus (-) quantity when the water level is below the impeller (suction lift), and a positive (+) quantity when the water is above (suction head).
H_f	=	Friction head and entrance losses in the suction piping at the system flow rate.
H_{vp}	=	Vapor pressure at the pumping water temperature.

Absolute Vapor Pressure		
Temperature °F	Feet	psi
50	0.411	0.178
60	0.591	0.256
70	0.839	0.363
80	1.171	0.507
90	1.612	0.698
100	2.192	0.949

Example: With a site altitude of 2,000 feet, water temperature of 60 degrees, suction assembly total head loss of 7.4 feet and the pump installed 5 feet above the water surface, determine the NPSHA:

H_a =	31.6 feet of head at 2,000 feet, derated 85 percent for possible weather changes	26.9	feet
H_s =	suction lift	- 5.0	feet
H_f =	friction losses	- 7.4	feet
H_{vp} =	vapor pressure (See the following table)	-0.591	feet
	NPSHA	13.909	feet

If the pump NPSHR is 10 feet, then the pump should perform properly with 13.91 feet of head. See Chapters 9 and 10 for information on calculating friction losses in pipes and fittings.

Velocity head should be calculated and added to the total suction assembly head loss.

Atmospheric Pressure at Varied Altitudes			
Altitude	Atmospheric Pressure		Barometric Pressure
in feet	feet of head	psi	inches Hg
Sea Level	34.0	14.7	29.9
500	33.4	14.4	29.4
1000	32.8	14.2	28.9
1500	32.1	13.9	28.4
2000	31.6	13.7	27.8
3000	30.5	13.2	26.8
4000	29.3	12.7	25.8
5000	28.2	12.2	24.8
6000	27.3	11.8	23.9
7000	26.1	11.3	23.0
8000	25.2	10.9	22.1
9000	24.3	10.5	21.3
10000	23.3	10.1	20.5

Velocity head is measured in feet of head and can be determined using the following formula:

$$H_v = \frac{V^2}{2g}$$

H_v = Velocity Head in feet
V = Velocity of the water flow in the pipe
g = Acceleration due to gravity, 32.2 feet per second

With a velocity of 5.0 feet per second the calculation is:

$$\frac{(5.0 \times 5.0 = 25.0)}{(2 \times 32.2 = 64.4)} = 0.39 \text{ Velocity head}$$

Several years ago, an irrigation system with a booster pump was designed and installed in a portion of a large cemetery. When the turf began turning brown it was discovered that the pump was operating at 75 percent of capacity and system runtimes were longer then anticipated. The system was designed to operate four zones simultaneously at 240 gpm, but the pump started cavitating when the fourth zone was turned on.

The system was connected to an existing 8-inch main at the end of a lake water pumping system, and an investigation revealed that the main was 1/2 to 2/3 full of debris resulting in flow restrictions, increased system friction and reduced pumping capacity. Consequently, a self-flushing filter system was installed at the upstream side of the booster pump to protect the entire system. The NSPHA was decreased below that required (NPSHR) by the pump resulting in cavitation at increased flow conditions.

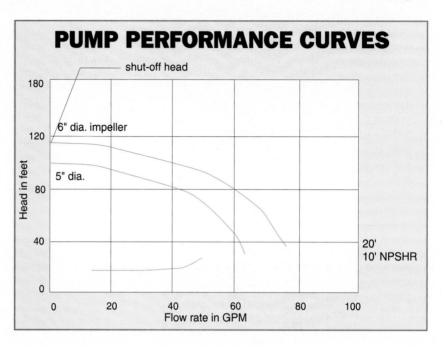

PUMP CURVES. A pump curve is used to indicate the flow rate of the pump at varying pressures, impeller sizes, NPSHR, brake horsepower requirements (BHP) and efficiency. The pump curve shown at the top of this page indicates two capacity curves (dark line), one curve is for a 6-inch diameter impeller, and the other a 5-inch diameter impeller. The flow rate for the 6-inch impeller ranges from 110 feet of head at zero flow, to 40 feet at 72 GPM, and the 5-inch diameter impeller shown below

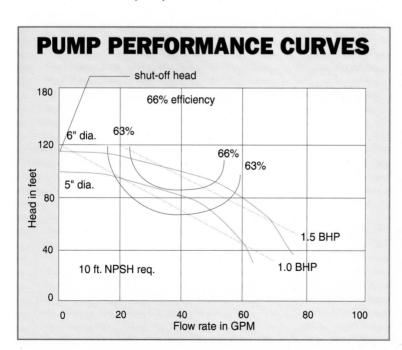

the 6-inch impeller indicates a range from 88 feet of head at zero flow and 37 feet of head at 60 GPM.

The pump curve on the bottom of page 162 shows additional information that will help in selecting the right pump, impeller, and horsepower. Pump efficiency changes as the pump flow rate varies. The efficiency curves are arcs (labeled 63 percent and 66 percent) which indicate the percentage of available water horsepower vs. the brake horsepower required to turn the pump impeller. The highest efficiency point on the 6-inch diameter impeller curve is at 38 gpm with 88 feet of head. The pump and impeller selection should be based on the system flow and pressure requirements, while maintaining high efficiencies.

The dotted lines on the chart indicate the brake horsepower requirements required to operate the pump at various flow and pressure conditions. A flow of 40 gpm at 70 feet of head would require a 1.0 BHP motor.

To calculate BHP use the following formula:

$$\frac{Q \times H}{3960 \times Ep}$$

Where:
BHP = Q = flow rate in gpm
H = head in feet
Ep = pump efficiency from pump curve (66% = 0.66)

Example: With a pump discharge of 40 gpm at 70 feet of head and 66 percent efficiency, what is the BHP?

$$\frac{40 \times 70 = 2800}{3960 \times 0.66 = 2613.6} = 1.07$$

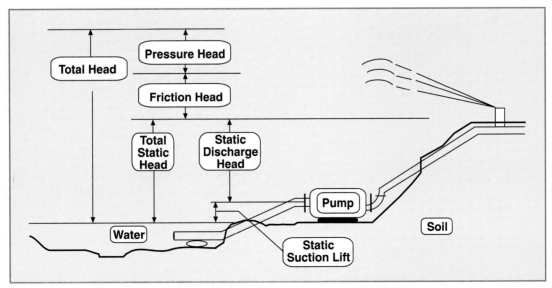

Static and operating heads when water level is below the centerline of the pump impeller.

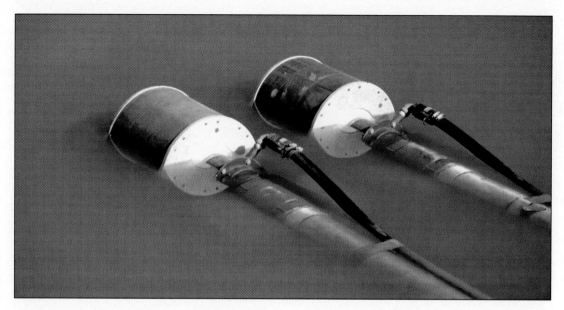

Pictured here are twin self-flushing intake sreens.

INTAKE & SUCTION LINE INSTALLATION. Avoid any air entrapment in the lines when installing the suction intake and piping. Always use an eccentric reducer coupling at the pump intake with the flat or straight side up, instead of the common concentric reducer fitting. This reduces the chance of air entrapment in the suction piping.

Always use 45 degree elbows for changes in direction and maintain a uniform incline to the pump when suction lift is a factor. Install a straight length of pipe that is a minimum of five pipe diameters in length between the eccentric reducer and the first 45 degree elbow to reduce turbulence at the pump inlet. If this is unsuitable, install a straightening vane upstream of the eccentric reducer.

Install a vacuum gauge with a test cock on the straight length of pipe upstream from the eccentric reducer. The vacuum gauge reads in inches of mercury (Hg) and can be used to verify the friction head loss in the intake piping. The multiplier to convert Hg to head in feet is 1.13. If the vacuum gauge reads 6.5 inches of Hg, multiply 1.13 x 6.5 to determine the total

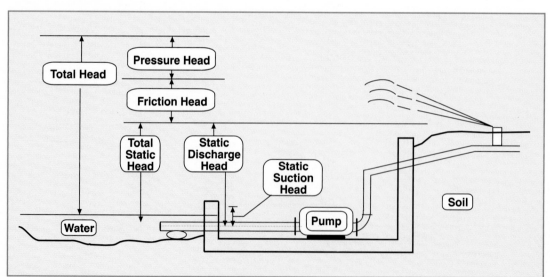

Static and operating heads when water level is above the centerline of the pump impeller.

suction head of 7.35 feet, then subtract the elevation difference between the pump impeller and the water level.

If the elevation difference is 5 feet, the actual friction head loss is 2.35 feet. Try keeping the friction head loss below 3 feet if possible. Compare the actual reading to original calculations to verify accuracy.

Water intakes from lakes, rivers and ditches should be located well below the surface of the water to prevent a vortex and air entering the intake screen and pipe. Also, keep the inlet (4-inches and smaller) at least 18 inches from the bottom to avoid debris intake that could be stirred up from the intake turbulence. Intake velocity through the screen should be kept to less than 0.5 feet per second. See Chapter 5 for additional information on screens and filtration.

PUMP & DISCHARGE FITTINGS. A pressure gauge, isolation valve, check valve and unions are recommended at the pump discharge. Additionally, air relief valves should be installed at all high points in the system to remove trapped air.

Install pipe supports for pipe sizes of 3 inches and larger. Install the pump on a raised concrete base with rubber mounts between the pump and the concrete.

Pressure and flow switches can be used to protect the pump if the water intake flow is interrupted. Pressure relief valves and pressure control valves can be installed to avoid excess pressure.

QUESTIONS:

1. What is suction lift?

2. What is the NPSHA with a site altitude of 1,000 feet, water temperature of 70,° suction assembly total head loss of 2.85 feet and the pump installed 9 feet above the water surface?

3. What is the absolute vapor pressure for water at 80 degrees?

4. What is the velocity head of water flowing at 4.5 feet per second?

5. What is the shutoff head?

6. What is the BHP requirements for a pump discharging 35 gpm at 82 feet of head and a 72 percent efficiency?

_____

7. Referring to the Pump Performance Curve on the bottom of page 162, what is the flow rate of the 6-inch impeller at 80 feet of head?

_____

(Answers to these questions are found on page 258.)

TABLES AND CONVERSION DATA

FRICTION LOSS AND VELOCITY

Pipe friction loss charts are based on the HAZEN AND WILLIAMS formula:

$$0.2083 \times \frac{100^{1.852}}{C^{1.852}} \times \frac{GPM^{1.852}}{D^{4.8655}} = F \times 0.433 = P$$

Where:

F = Friction loss in feet of head
C = Friction factor for smoothness
D = Inside diameter in inches
P = Friction loss in psi

Velocity charts are based on the following formula:

$$\frac{GPM}{2.45 \times ID\ squared} = Velocity\ (fps)$$

Where:

ID = Inside diameter in inches

HYDRAULIC CALCULATIONS

Project: _____   Date: _____ By: _____

Min. Water Pressure: _____   Obtained from: _____ At: _____

Valve No. _____	Valve No. _____
Flow _____	Flow _____

Pressure Available:

Min. pressure at tap in main:	_____	Min. pressure at tap in main:	_____
Regulated pressure:	_____	Regulated pressure:	_____
Gain due to elevation:	_____	Gain due to elevation:	_____
Total available pressure:	_____	Total available pressure:	_____

Pressure Losses:	Size	GPM	Feet*	Vel f/s	Losses
Losses due to Elev.					
Service Line					
Meter					
Backflow Preventer					
Control Valve					
Mainline					
Mainline					
Mainline					
Mainline					
Mainline					
Lateral Line					
Lateral Line					
Lateral Line					
Lateral Line					
Lateral Line					
Miscellaneous					
Miscellaneous					
Miscellaneous					

Pressure Losses:	Size	GPM	Feet*	Vel f/s	Losses
Losses due to Elev.					
Service Line					
Meter					
Backflow Preventer					
Control Valve					
Mainline					
Mainline					
Mainline					
Mainline					
Mainline					
Lateral Line					
Lateral Line					
Lateral Line					
Lateral Line					
Lateral Line					
Miscellaneous					
Miscellaneous					
Miscellaneous					

Total Pressure Loss:	_____	Total Pressure Loss:	_____
Pressure at Head:	[____]	Pressure at Head:	[____]
Min. required at head:	_____	Min. required at head:	_____

Length of pipe includes equivalent length for fittings.

STANDARD PIPE DIMENSIONS (inches)

Plastic Pipe

Pipe Size	O.D.	SDR 26 (CL 160)		SDR 21 (CL 200)		SDR 13.5 (CL 315)	
		I.D.	Wall Thickness	I.D.	Wall Thickness	I.D.	Wall Thickness
3/4"	1.05	-	-	0.93	0.06	0.894	0.078
1"	1.315	1.195	0.06	1.189	0.063	1.121	0.097
1 1/4"	1.66	1.532	0.064	1.502	0.079	1.414	0.123
1 1/2"	1.9	1.754	0.073	1.72	0.09	1.618	0.141
2"	2.375	2.193	0.091	2.149	0.113	2.023	0.176
2 1/2"	2.875	2.655	0.11	2.601	0.137	2.449	0.213
3"	3.5	3.23	0.135	3.166	0.167	2.982	0.259
4"	4.5	4.154	0.173	4.072	0.214	3.834	0.333
6"	6.625	6.115	0.255	5.993	0.316	5.643	0.491
8"	8.625	7.961	0.332	7.805	0.41	-	-
10"	10.75	9.924	0.413	9.728	0.511	-	-
12"	12.75	11.77	0.49	11.538	0.606	-	-

Pipe Size	O.D.	SCH. 40		SCH. 80	
		I.D.	Wall Thickness	I.D.	Wall Thickness
3/4"	1.05	0.824	0.113	0.742	0.154
1"	1.315	1.049	0.133	0.957	0.179
1 1/4"	1.66	1.38	0.14	1.278	0.191
1 1/2"	1.9	1.61	0.145	1.5	0.2
2"	2.375	2.067	0.154	1.939	0.218
2 1/2"	2.875	2.469	0.203	2.323	0.276
3"	3.5	3.068	0.216	2.9	0.3
4"	4.5	4.026	0.237	3.826	0.337

Pipe Size	I.D.	SDR 15 / 80#		SDR 11.5 / 100# NSF	
		O.D.	Wall Thickness	O.D.	Wall Thickness
1/2"	0.622	0.742	0.06	0.742	0.06
3/4"	0.824	0.944	0.06	0.968	0.072
1"	1.049	1.189	0.07	1.231	0.091
1 1/4"	1.38	1.564	0.092	1.62	0.12
1 1/2"	1.61	1.824	0.107	1.89	0.14
2"	2.067	2.343	0.138	2.427	0.18
2 1/2"	2.469	2.799	0.165	2.899	0.215

Pipe Size	O.D.	Type M		Type L		Type K	
		I.D.	Wall Thickness	I.D.	Wall Thickness	I.D.	Wall Thickness
3/4"	0.875	0.811	0.032	0.785	0.045	0.745	0.065
1"	1.125	1.055	0.035	1.025	0.05	0.995	0.065
1 1/4"	1.375	1.291	0.042	1.265	0.055	1.245	0.065
1 1/2"	1.625	1.527	0.049	1.505	0.06	1.481	0.072
2"	2.125	2.009	0.058	1.985	0.07	1.959	0.083
2 1/2"	2.625	2.495	0.065	2.465	0.08	2.435	0.095
3"	3.125	2.981	0.072	2.945	0.09	2.907	0.109
4"	4.125	3.935	0.095	3.905	0.11	3.857	0.134

PRESSURE LOSS PER 100 OF SDR 26 / CL 160 PLASTIC PIPE (lbs./sq.in.)

C=150

GPM	1"	1 1/4"	1 1/2"	2"	2 1/2"	3"	4"
1	0.018	0.005	0.003	0.001			
2	0.065	0.019	0.010	0.003	0.001		
3	0.137	0.041	0.021	0.007	0.003	0.001	
4	0.233	0.070	0.036	0.012	0.005	0.002	
5	0.352	0.105	0.054	0.018	0.007	0.003	
6	0.494	0.148	0.076	0.026	0.010	0.004	0.001
7	0.657	0.196	0.102	0.034	0.014	0.005	0.002
8	0.842	0.251	0.130	0.044	0.017	0.007	0.002
9	1.047	0.313	0.162	0.055	0.022	0.008	0.002
10	1.272	0.380	0.197	0.066	0.026	0.010	0.003
11	1.518	0.453	0.235	0.079	0.031	0.012	0.004
12	1.783	0.533	0.276	0.093	0.037	0.014	0.004
13	2.068	0.618	0.320	0.108	0.043	0.016	0.005
14	2.373	0.708	0.367	0.124	0.049	0.019	0.006
15	2.696	0.805	0.417	0.141	0.055	0.021	0.006
20	4.593	1.371	0.710	0.239	0.094	0.036	0.011
25	6.944	2.073	1.073	0.362	0.143	0.055	0.016
30	9.733	2.906	1.504	0.507	0.200	0.077	0.023
35	12.949	3.866	2.001	0.675	0.266	0.103	0.030
40		4.951	2.563	0.864	0.341	0.131	0.039
50		7.485	3.875	1.307	0.516	0.199	0.058
60		10.491	5.431	1.832	0.723	0.278	0.082
70		13.957	7.225	2.437	0.961	0.370	0.109
80			9.252	3.121	1.231	0.474	0.139
90			11.508	3.881	1.531	0.590	0.173
100			13.987	4.718	1.861	0.717	0.211
120				6.613	2.609	1.005	0.295
140				8.798	3.471	1.337	0.393
160				11.266	4.444	1.712	0.503
180					5.528	2.130	0.626
200					6.719	2.588	0.761
220					8.016	3.088	0.908
240					9.417	3.628	1.067
260					10.922	4.208	1.237
280					12.529	4.827	1.419
300						5.485	1.613

Shaded areas are velocities over 5 ft./sec. Use with caution.

*** For 3/4" pipe, see CL 315 table.**

VELOCITY OF FLOW FOR SDR 26 / CL 160 PLASTIC PIPE (ft./sec.)

GPM	1"	1 1/4"	1 1/2"	2"	2 1/2"	3"	4"
1	0.286	0.174	0.133	0.085	0.058	0.039	0.024
2	0.572	0.348	0.265	0.170	0.116	0.078	0.047
3	0.857	0.522	0.398	0.255	0.174	0.117	0.071
4	1.143	0.696	0.531	0.339	0.232	0.156	0.095
5	1.429	0.870	0.663	0.424	0.290	0.196	0.118
6	1.715	1.043	0.796	0.509	0.347	0.235	0.142
7	2.001	1.217	0.929	0.594	0.405	0.274	0.166
8	2.287	1.391	1.061	0.679	0.463	0.313	0.189
9	2.572	1.565	1.194	0.764	0.521	0.352	0.213
10	2.858	1.739	1.327	0.849	0.579	0.391	0.237
11	3.144	1.913	1.459	0.934	0.637	0.430	0.260
12	3.430	2.087	1.592	1.018	0.695	0.469	0.284
13	3.716	2.261	1.725	1.103	0.753	0.509	0.307
14	4.002	2.435	1.857	1.188	0.811	0.548	0.331
15	4.287	2.609	1.990	1.273	0.869	0.587	0.355
20	5.716	3.478	2.653	1.697	1.158	0.782	0.473
25	7.146	4.348	3.317	2.122	1.448	0.978	0.591
30	8.575	5.217	3.980	2.546	1.737	1.174	0.710
35	10.004	6.087	4.643	2.970	2.027	1.369	0.828
40		6.956	5.307	3.395	2.316	1.565	0.946
50		8.695	6.634	4.244	2.895	1.956	1.183
60		10.434	7.960	5.092	3.474	2.347	1.419
70		12.173	9.287	5.941	4.053	2.739	1.656
80			10.614	6.790	4.632	3.130	1.892
90			11.940	7.638	5.211	3.521	2.129
100			13.267	8.487	5.790	3.912	2.365
120				10.184	6.948	4.695	2.838
140				11.882	8.106	5.477	3.312
160				13.579	9.265	6.260	3.785
180					10.423	7.042	4.258
200					11.581	7.825	4.731
220					12.739	8.607	5.204
240					13.897	9.389	5.677
260					15.055	10.172	6.150
280					16.213	10.954	6.623
300						11.737	7.096

* Shaded areas are velocities over 5 ft./sec. Use with caution.
** For 3/4" pipe, see CL 315 table.

PRESSURE LOSS PER 100' OF SDR 21 / CL 200 PLASTIC PIPE (lbs./sq. in.)

C=150

GPM	3/4"	1"	1 1/4"	1 1/2"	2"	2 1/2"
1	0.061	0.018	0.006	0.003	0.001	0.000
2	0.219	0.066	0.021	0.011	0.004	0.001
3	0.463	0.140	0.045	0.023	0.008	0.003
4	0.790	0.239	0.077	0.040	0.013	0.005
5	1.194	0.361	0.116	0.060	0.020	0.008
6	1.673	0.506	0.162	0.084	0.028	0.011
7	2.226	0.674	0.216	0.112	0.038	0.015
8	2.851	0.863	0.277	0.143	0.048	0.019
9	3.545	1.073	0.344	0.178	0.060	0.024
10	4.309	1.304	0.418	0.216	0.073	0.029
11	5.141	1.556	0.499	0.258	0.087	0.035
12	6.040	1.828	0.586	0.303	0.103	0.041
13	7.005	2.120	0.680	0.352	0.119	0.047
14	8.036	2.432	0.780	0.403	0.137	0.054
15	9.131	2.763	0.886	0.458	0.155	0.061
20		4.707	1.510	0.781	0.264	0.104
25		7.116	2.283	1.181	0.400	0.158
30		9.974	3.200	1.655	0.560	0.221
35			4.257	2.201	0.745	0.294
40			5.451	2.819	0.954	0.377
50			8.241	4.262	1.442	0.570
60				5.973	2.022	0.799
70				7.947	2.690	1.062
80				10.177	3.444	1.361
90					4.284	1.692
100					5.207	2.057
110					6.212	2.454
120					7.298	2.883
130					8.464	3.344
140					9.709	3.835
150						4.358

Shaded areas are velocities over 5 ft./sec. Use with caution.

VELOCITY OF FLOW FOR SDR 21 / CL 200 PLASTIC PIPE (ft./sec.)

GPM	3/4"	1"	1 1/4"	1 1/2"	2"	2 1/2"
1	0.472	0.289	0.181	0.138	0.088	0.060
2	0.944	0.577	0.362	0.276	0.177	0.121
3	1.416	0.866	0.543	0.414	0.265	0.181
4	1.888	1.155	0.724	0.552	0.354	0.241
5	2.360	1.444	0.905	0.690	0.442	0.302
6	2.832	1.732	1.086	0.828	0.530	0.362
7	3.303	2.021	1.266	0.966	0.619	0.422
8	3.775	2.310	1.447	1.104	0.707	0.483
9	4.247	2.598	1.628	1.242	0.795	0.543
10	4.719	2.887	1.809	1.380	0.884	0.603
11	5.191	3.176	1.990	1.518	0.972	0.664
12	5.663	3.465	2.171	1.656	1.061	0.724
13	6.135	3.753	2.352	1.794	1.149	0.784
14	6.607	4.042	2.533	1.932	1.237	0.845
15	7.079	4.331	2.714	2.070	1.326	0.905
20	9.438	5.774	3.618	2.759	1.768	1.207
25	11.798	7.218	4.523	3.449	2.210	1.508
30		8.661	5.428	4.139	2.651	1.810
35		10.105	6.332	4.829	3.093	2.112
40		11.549	7.237	5.519	3.535	2.413
50			9.046	6.898	4.419	3.017
60			10.855	8.278	5.303	3.620
70			12.665	9.658	6.187	4.223
80				11.037	7.071	4.827
90				12.417	7.954	5.430
100				13.797	8.838	6.033
110					9.722	6.637
120					10.606	7.240
130					11.490	7.843
140					12.373	8.447
150					13.257	9.050

Shaded areas are velocities over 5 ft./sec. Use with caution.

PRESSURE LOSS PER 100' OF SDR 21 / CL 200 PLASTIC PIPE (lbs./sq. in.)

C=150

GPM	3"	4"	6"	8"	10"	12"
50	0.219	0.064	0.010	0.003	0.001	
75	0.464	0.136	0.021	0.006	0.002	0.001
100	0.790	0.232	0.035	0.010	0.003	0.001
125	1.195	0.351	0.054	0.015	0.005	0.002
150	1.675	0.492	0.075	0.021	0.007	0.003
175	2.228	0.655	0.100	0.028	0.009	0.004
200	2.853	0.839	0.128	0.035	0.012	0.005
250	4.313	1.268	0.193	0.053	0.018	0.008
300	6.046	1.777	0.271	0.075	0.026	0.011
350	8.043	2.364	0.361	0.100	0.034	0.015
400	10.300	3.027	0.462	0.128	0.044	0.019
450	12.811	3.765	0.574	0.159	0.054	0.024
500	15.571	4.576	0.698	0.193	0.066	0.029
600	21.825	6.415	0.979	0.271	0.093	0.040
700	29.036	8.534	1.302	0.360	0.123	0.054
800	37.183	10.928	1.667	0.461	0.158	0.069
900	46.246	13.592	2.073	0.573	0.196	0.086
1000	56.211	16.521	2.520	0.697	0.239	0.104
1500		35.007	5.340	1.477	0.506	0.220
2000		59.641	9.098	2.516	0.862	0.376
2500			13.753	3.804	1.303	0.568
3000			19.278	5.331	1.826	0.796
3500			25.647	7.093	2.429	1.059
4000			32.843	9.083	3.111	1.356
4500				11.297	3.869	1.687
5000				13.731	4.702	2.050
6000				19.247	6.591	2.873
7000				25.606	8.769	3.823
8000				32.790	11.229	4.895
9000				40.782	13.966	6.089
10000					16.976	7.400

Shaded areas are velocities over 5 ft./sec. Use with caution.

VELOCITY OF FLOW FOR SDR 21 / CL 200 PLASTIC PIPE (ft. sec.)

GPM	3"	4"	6"	8"	10"	12"
50	2.036	1.231	0.568	0.335	0.216	0.153
75	3.054	1.846	0.852	0.503	0.323	0.230
100	4.072	2.462	1.136	0.670	0.431	0.307
125	5.090	3.077	1.421	0.838	0.539	0.383
150	6.108	3.692	1.705	1.005	0.647	0.460
175	7.126	4.308	1.989	1.173	0.755	0.537
200	8.144	4.923	2.273	1.340	0.863	0.613
250	10.180	6.154	2.841	1.675	1.078	0.767
300	12.216	7.385	3.409	2.010	1.294	0.920
350	14.252	8.616	3.978	2.345	1.510	1.073
400	16.288	9.846	4.546	2.680	1.725	1.226
450	18.324	11.077	5.114	3.015	1.941	1.380
500	20.360	12.308	5.682	3.350	2.157	1.533
600	24.432	14.770	6.819	4.020	2.588	1.840
700	28.504	17.231	7.955	4.690	3.019	2.146
800	32.576	19.693	9.091	5.360	3.450	2.453
900	36.648	22.154	10.228	6.030	3.882	2.759
1000	40.720	24.616	11.364	6.700	4.313	3.066
1500		36.924	17.047	10.050	6.470	4.599
2000		49.232	22.729	13.400	8.626	6.132
2500			28.411	16.751	10.783	7.665
3000			34.093	20.101	12.939	9.198
3500			39.775	23.451	15.096	10.731
4000			45.457	26.801	17.252	12.264
4500				30.151	19.409	13.797
5000				33.501	21.565	15.330
6000				40.201	25.878	18.396
7000				46.901	30.192	21.462
8000				53.602	34.505	24.528
9000				60.302	38.818	27.594
10000					43.131	30.660

* Shaded areas are velocities over 5 ft./sec. Use with caution.

PRESSURE LOSS PER 100' OF SDR 13.5 / CL 315 PLASTIC PIPE (lbs./sq. in.)

C=150

GPM	3/4"	1"	1 1/4"	1 1/2"	2"	2 1/2"	3"	4"
1	0.073	0.024	0.008	0.004	0.001	0.001		
2	0.265	0.088	0.028	0.015	0.005	0.002	0.001	
3	0.562	0.187	0.060	0.031	0.011	0.004	0.002	
4	0.957	0.318	0.103	0.053	0.018	0.007	0.003	0.001
5	1.446	0.481	0.155	0.081	0.027	0.011	0.004	0.001
6	2.027	0.674	0.218	0.113	0.038	0.015	0.006	0.002
7	2.697	0.897	0.290	0.150	0.051	0.020	0.008	0.002
8	3.454	1.149	0.371	0.193	0.065	0.026	0.010	0.003
9	4.296	1.429	0.462	0.240	0.081	0.032	0.012	0.004
10	5.222	1.737	0.561	0.291	0.098	0.039	0.015	0.004
11	6.230	2.072	0.669	0.347	0.117	0.046	0.018	0.005
12	7.319	2.434	0.786	0.408	0.138	0.054	0.021	0.006
13	8.489	2.823	0.912	0.473	0.160	0.063	0.024	0.007
14	9.738	3.238	1.046	0.543	0.183	0.072	0.028	0.008
15		3.680	1.189	0.617	0.208	0.082	0.032	0.009
20		6.269	2.026	1.051	0.355	0.140	0.054	0.016
25		9.477	3.062	1.589	0.536	0.212	0.081	0.024
30			4.292	2.228	0.751	0.297	0.114	0.033
35			5.710	2.964	1.000	0.394	0.151	0.045
40			7.312	3.796	1.280	0.505	0.194	0.057
50				5.738	1.935	0.764	0.293	0.086
60				8.043	2.712	1.070	0.411	0.121
70					3.609	1.424	0.546	0.161
80					4.621	1.824	0.700	0.206
90					5.748	2.268	0.870	0.256
100					6.986	2.757	1.058	0.311
120					9.792	3.864	1.482	0.436
140						5.141	1.972	0.581
160						6.584	2.526	0.744
180						8.188	3.141	0.925
200						9.953	3.818	1.124
220							4.555	1.341
240							5.352	1.576
260							6.207	1.827
280							7.120	2.096
300							8.090	2.382

Shaded areas are velocities over 5 ft./sec. Use with caution.

VELOCITY OF FLOW FOR SDR 13.5 / CL 315 PLASTIC PIPE (ft./sec.)

GPM	3/4"	1"	1 1/4"	1 1/2"	2"	2 1/2"	3"	4"
1	0.511	0.325	0.204	0.156	0.100	0.068	0.046	0.028
2	1.021	0.650	0.408	0.312	0.199	0.136	0.092	0.056
3	1.532	0.974	0.612	0.468	0.299	0.204	0.138	0.083
4	2.043	1.299	0.817	0.624	0.399	0.272	0.184	0.111
5	2.553	1.624	1.021	0.780	0.499	0.340	0.230	0.139
6	3.064	1.949	1.225	0.935	0.598	0.408	0.275	0.167
7	3.575	2.274	1.429	1.091	0.698	0.476	0.321	0.194
8	4.086	2.598	1.633	1.247	0.798	0.544	0.367	0.222
9	4.596	2.923	1.837	1.403	0.898	0.612	0.413	0.250
10	5.107	3.248	2.041	1.559	0.997	0.681	0.459	0.278
11	5.618	3.573	2.246	1.715	1.097	0.749	0.505	0.305
12	6.128	3.898	2.450	1.871	1.197	0.817	0.551	0.333
13	6.639	4.222	2.654	2.027	1.297	0.885	0.597	0.361
14	7.150	4.547	2.858	2.183	1.396	0.953	0.643	0.389
15	7.660	4.872	3.062	2.339	1.496	1.021	0.689	0.417
20	10.214	6.496	4.083	3.118	1.995	1.361	0.918	0.555
25	12.767	8.120	5.104	3.898	2.493	1.701	1.148	0.694
30		9.744	6.124	4.677	2.992	2.042	1.377	0.833
35		11.368	7.145	5.457	3.491	2.382	1.607	0.972
40		12.992	8.166	6.236	3.989	2.722	1.836	1.111
50			10.207	7.796	4.987	3.403	2.295	1.388
60			12.249	9.355	5.984	4.083	2.754	1.666
70				10.914	6.981	4.764	3.213	1.944
80				12.473	7.979	5.444	3.672	2.221
90					8.976	6.125	4.131	2.499
100					9.973	6.805	4.590	2.777
120					11.968	8.167	5.508	3.332
140					13.963	9.528	6.426	3.887
160						10.889	7.344	4.443
180						12.250	8.262	4.998
200						13.611	9.180	5.553
220							10.098	6.109
240							11.016	6.664
260							11.934	7.219
280							12.852	7.775
300							13.770	8.330

Shaded areas are velocities over 5 ft./sec. Use with caution.

PRESSURE LOSS PER 100' OF SCHEDULE 40 PLASTIC PIPE (lbs./sq. in.)

C=150

GPM	3/4"	1"	1 1/4"	1 1/2"	2"	2 1/2"	3"	4"
1	0.109	0.034	0.009	0.004	0.001	0.001		
2	0.394	0.122	0.032	0.015	0.004	0.002	0.001	
3	0.835	0.258	0.068	0.032	0.010	0.004	0.001	
4	1.423	0.440	0.116	0.055	0.016	0.007	0.002	
5	2.151	0.664	0.175	0.083	0.025	0.010	0.004	0.001
6	3.015	0.931	0.245	0.116	0.034	0.014	0.005	0.001
7	4.011	1.239	0.326	0.154	0.046	0.019	0.007	0.002
8	5.136	1.587	0.418	0.197	0.059	0.025	0.009	0.002
9	6.388	1.973	0.520	0.245	0.073	0.031	0.011	0.003
10	7.764	2.399	0.632	0.298	0.088	0.037	0.013	0.003
11	9.263	2.862	0.754	0.356	0.106	0.044	0.015	0.004
12	10.883	3.362	0.885	0.418	0.124	0.052	0.018	0.005
13	12.622	3.899	1.027	0.485	0.144	0.061	0.021	0.006
14	14.479	4.473	1.178	0.556	0.165	0.069	0.024	0.006
15	16.453	5.083	1.338	0.632	0.187	0.079	0.027	0.007
20	28.030	8.659	2.280	1.077	0.319	0.135	0.047	0.012
25	42.374	13.091	3.447	1.628	0.483	0.203	0.071	0.019
30	59.394	18.349	4.832	2.282	0.677	0.285	0.099	0.026
35		24.411	6.428	3.036	0.900	0.379	0.132	0.035
40		31.260	8.232	3.888	1.153	0.486	0.169	0.045
50		47.257	12.444	5.878	1.743	0.734	0.255	0.068
60			17.443	8.239	2.443	1.029	0.358	0.095
70			23.206	10.961	3.250	1.369	0.476	0.127
80			29.716	14.037	4.162	1.753	0.609	0.162
90			36.960	17.458	5.176	2.180	0.758	0.202
100				21.220	6.292	2.650	0.921	0.245
110				25.317	7.506	3.162	1.099	0.293
120				29.743	8.819	3.714	1.291	0.344
130					10.228	4.308	1.497	0.399
140					11.733	4.942	1.717	0.458
150					13.332	5.615	1.952	0.520
160					15.024	6.328	2.199	0.586
170					16.810	7.080	2.461	0.656
180					18.687	7.871	2.735	0.729
190					20.655	8.700	3.024	0.806
200					22.713	9.566	3.325	0.886
210						10.471	3.639	0.970
220						11.413	3.967	1.057
230						12.393	4.307	1.148
240						13.409	4.660	1.242

Shaded areas are velocities over 5 ft./sec. Use with caution.

VELOCITY OF FLOW FOR SCHEDULE 40 PLASTIC PIPE (ft./sec.)

GPM	3/4"	1"	1 1/4"	1 1/2"	2"	2 1/2"	3"	4"
1	0.601	0.371	0.214	0.157	0.096	0.067	0.043	0.025
2	1.202	0.742	0.429	0.315	0.191	0.134	0.087	0.050
3	1.803	1.113	0.643	0.472	0.287	0.201	0.130	0.076
4	2.405	1.484	0.857	0.630	0.382	0.268	0.173	0.101
5	3.006	1.855	1.072	0.787	0.478	0.335	0.217	0.126
6	3.607	2.226	1.286	0.945	0.573	0.402	0.260	0.151
7	4.208	2.596	1.500	1.102	0.669	0.469	0.304	0.176
8	4.809	2.967	1.715	1.260	0.764	0.536	0.347	0.201
9	5.410	3.338	1.929	1.417	0.860	0.603	0.390	0.227
10	6.011	3.709	2.143	1.575	0.955	0.670	0.434	0.252
11	6.613	4.080	2.358	1.732	1.051	0.737	0.477	0.277
12	7.214	4.451	2.572	1.890	1.146	0.803	0.520	0.302
13	7.815	4.822	2.786	2.047	1.242	0.870	0.564	0.327
14	8.416	5.193	3.001	2.205	1.337	0.937	0.607	0.353
15	9.017	5.564	3.215	2.362	1.433	1.004	0.650	0.378
20	12.023	7.418	4.287	3.149	1.911	1.339	0.867	0.504
25	15.029	9.273	5.358	3.937	2.388	1.674	1.084	0.630
30	18.034	11.128	6.430	4.724	2.866	2.009	1.301	0.755
35		12.982	7.501	5.511	3.344	2.343	1.518	0.881
40		14.837	8.573	6.299	3.821	2.678	1.735	1.007
50		18.546	10.716	7.873	4.777	3.348	2.168	1.259
60			12.860	9.448	5.732	4.017	2.602	1.511
70			15.003	11.023	6.687	4.687	3.035	1.763
80			17.146	12.597	7.643	5.357	3.469	2.015
90			19.289	14.172	8.598	6.026	3.903	2.266
100				15.746	9.553	6.696	4.336	2.518
110				17.321	10.509	7.365	4.770	2.770
120				18.896	11.464	8.035	5.204	3.022
130					12.419	8.704	5.637	3.274
140					13.375	9.374	6.071	3.525
150					14.330	10.043	6.505	3.777
160					15.285	10.713	6.938	4.029
170					16.241	11.383	7.372	4.281
180					17.196	12.052	7.805	4.533
190					18.151	12.722	8.239	4.785
200					19.107	13.391	8.673	5.036
210						14.061	9.106	5.288
220						14.730	9.540	5.540
230						15.400	9.974	5.792
240						16.070	10.407	6.044

* Shaded areas are velocities over 5 ft./sec. Use with caution.

PRESSURE LOSS PER 100' OF SCHEDULE 80
PLASTIC PIPE (lbs./sq. in.)

C=150

GPM	3/4"	1"	1 1/4"	1 1/2"	2"	2 1/2"	3"	4"
1	0.182	0.053	0.013	0.006	0.002	0.001		
2	0.656	0.190	0.047	0.021	0.006	0.003	0.001	
3	1.391	0.403	0.099	0.045	0.013	0.005	0.002	
4	2.369	0.687	0.168	0.077	0.022	0.009	0.003	
5	3.582	1.039	0.254	0.117	0.033	0.014	0.005	0.001
6	5.020	1.456	0.356	0.163	0.047	0.019	0.007	0.002
7	6.679	1.937	0.474	0.217	0.062	0.026	0.009	0.002
8	8.553	2.480	0.607	0.278	0.080	0.033	0.011	0.003
9	10.638	3.085	0.755	0.346	0.099	0.041	0.014	0.004
10	12.930	3.749	0.918	0.421	0.121	0.050	0.017	0.004
11		4.473	1.095	0.502	0.144	0.060	0.020	0.005
12		5.255	1.286	0.590	0.169	0.070	0.024	0.006
13		6.095	1.492	0.684	0.196	0.081	0.028	0.007
14		6.991	1.711	0.785	0.225	0.093	0.032	0.008
15		7.944	1.945	0.892	0.256	0.106	0.036	0.009
20		13.534	3.313	1.520	0.436	0.181	0.061	0.016
25			5.009	2.298	0.659	0.274	0.093	0.024
30			7.020	3.220	0.924	0.383	0.130	0.034
35			9.340	4.284	1.229	0.510	0.173	0.045
40			11.960	5.487	1.573	0.653	0.222	0.058
50				8.294	2.379	0.987	0.336	0.087
60				11.626	3.334	1.384	0.470	0.122
70					4.436	1.841	0.626	0.162
80					5.680	2.358	0.801	0.208
90					7.065	2.933	0.997	0.259
100					8.587	3.565	1.211	0.315
110					10.245	4.253	1.445	0.375
120					12.036	4.997	1.698	0.441
130					13.959	5.795	1.969	0.511
140						6.648	2.259	0.587
150						7.554	2.567	0.667
160						8.513	2.893	0.751
170						9.524	3.236	0.840
180						10.588	3.598	0.934
190						11.703	3.977	1.033
200						12.869	4.373	1.136
210							4.786	1.243
220							5.217	1.355
230							5.665	1.471
240							6.129	1.592

* Shaded areas are velocities over 5 ft./sec. Use with caution.

VELOCITY OF FLOW FOR SCHEDULE 80 PLASTIC PIPE (FT./SEC.)

GPM	3/4"	1"	1 1/4"	1 1/2"	2"	2 1/2"	3"	4"
1	0.741	0.446	0.250	0.181	0.109	0.076	0.049	0.028
2	1.483	0.891	0.500	0.363	0.217	0.151	0.097	0.056
3	2.224	1.337	0.750	0.544	0.326	0.227	0.146	0.084
4	2.965	1.783	1.000	0.726	0.434	0.303	0.194	0.112
5	3.707	2.228	1.250	0.907	0.543	0.378	0.243	0.139
6	4.448	2.674	1.499	1.088	0.651	0.454	0.291	0.167
7	5.189	3.120	1.749	1.270	0.760	0.529	0.340	0.195
8	5.931	3.565	1.999	1.451	0.868	0.605	0.388	0.223
9	6.672	4.011	2.249	1.633	0.977	0.681	0.437	0.251
10	7.414	4.457	2.499	1.814	1.086	0.756	0.485	0.279
11	8.155	4.902	2.749	1.995	1.194	0.832	0.534	0.307
12	8.896	5.348	2.999	2.177	1.303	0.908	0.582	0.335
13	9.638	5.794	3.249	2.358	1.411	0.983	0.631	0.362
14	10.379	6.239	3.499	2.540	1.520	1.059	0.679	0.390
15	11.120	6.685	3.749	2.721	1.628	1.135	0.728	0.418
20		8.913	4.998	3.628	2.171	1.513	0.971	0.558
25		11.142	6.248	4.535	2.714	1.891	1.213	0.697
30		13.370	7.497	5.442	3.257	2.269	1.456	0.836
35			8.747	6.349	3.800	2.647	1.699	0.976
40			9.996	7.256	4.342	3.025	1.941	1.115
50			12.495	9.070	5.428	3.782	2.427	1.394
60				10.884	6.514	4.538	2.912	1.673
70				12.698	7.599	5.295	3.397	1.952
80					8.685	6.051	3.883	2.231
90					9.771	6.807	4.368	2.509
100					10.856	7.564	4.853	2.788
110					11.942	8.320	5.339	3.067
120					13.027	9.076	5.824	3.346
130						9.833	6.309	3.625
140						10.589	6.795	3.904
150						11.346	7.280	4.182
160						12.102	7.765	4.461
170						12.858	8.251	4.740
180						13.615	8.736	5.019
190							9.221	5.298
200							9.707	5.577
210							10.192	5.855
220							10.677	6.134
230							11.163	6.413
240							11.648	6.692

Shaded areas are velocities over 5 ft./sec. Use with caution.

PRESSURE LOSS PER 100' OF POLYETHYLENE PIPE (lbs./sq. in.)

C=140

GPM	1/2"	3/4"	1"	1 1/4"	1 1/2"	2"	2 1/2"
1	0.487	0.124	0.038	0.010	0.005	0.001	0.001
2	1.759	0.448	0.138	0.036	0.017	0.005	0.002
3	3.728	0.949	0.293	0.077	0.036	0.011	0.005
4	6.351	1.617	0.499	0.132	0.062	0.018	0.008
5	9.602	2.444	0.755	0.199	0.094	0.028	0.012
6	13.459	3.426	1.058	0.279	0.132	0.039	0.016
7		4.557	1.408	0.371	0.175	0.052	0.022
8		5.836	1.803	0.475	0.224	0.066	0.028
9		7.259	2.242	0.591	0.279	0.083	0.035
10		8.823	2.726	0.718	0.339	0.101	0.042
11		10.526	3.252	0.856	0.404	0.120	0.051
12		12.367	3.820	1.006	0.475	0.141	0.059
13			4.431	1.167	0.551	0.163	0.069
14			5.083	1.338	0.632	0.187	0.079
15			5.775	1.521	0.718	0.213	0.090
20			9.839	2.591	1.224	0.363	0.153
25				3.917	1.850	0.549	0.231
30				5.490	2.593	0.769	0.324
35				7.304	3.450	1.023	0.431
40				9.354	4.418	1.310	0.552
50					6.679	1.980	0.834
60					9.362	2.776	1.169
70					12.455	3.693	1.555
80						4.729	1.992
90						5.882	2.477
100						7.149	3.011

Shaded areas are velocities over 5 ft./sec. Use with caution.

VELOCITY OF FLOW FOR POLYETHYLENE PIPE
(ft./sec.)

GPM	1/2"	3/4"	1"	1 1/4"	1 1/2"	2"	2 1/2"
1	1.055	0.601	0.371	0.214	0.157	0.096	0.067
2	2.110	1.202	0.742	0.429	0.315	0.191	0.134
3	3.165	1.803	1.113	0.643	0.472	0.287	0.201
4	4.220	2.405	1.484	0.857	0.630	0.382	0.268
5	5.275	3.006	1.855	1.072	0.787	0.478	0.335
6	6.330	3.607	2.226	1.286	0.945	0.573	0.402
7	7.385	4.208	2.596	1.500	1.102	0.669	0.469
8	8.440	4.809	2.967	1.715	1.260	0.764	0.536
9	9.495	5.410	3.338	1.929	1.417	0.860	0.603
10	10.550	6.011	3.709	2.143	1.575	0.955	0.670
11	11.605	6.613	4.080	2.358	1.732	1.051	0.737
12	12.660	7.214	4.451	2.572	1.890	1.146	0.803
13	13.715	7.815	4.822	2.786	2.047	1.242	0.870
14		8.416	5.193	3.001	2.205	1.337	0.937
15		9.017	5.564	3.215	2.362	1.433	1.004
20		12.023	7.418	4.287	3.149	1.911	1.339
25			9.273	5.358	3.937	2.388	1.674
30			11.128	6.430	4.724	2.866	2.009
35			12.982	7.501	5.511	3.344	2.343
40				8.573	6.299	3.821	2.678
50				10.716	7.873	4.777	3.348
60				12.860	9.448	5.732	4.017
70					11.023	6.687	4.687
80					12.597	7.643	5.357
90						8.598	6.026
100						9.553	6.696

* Shaded areas are velocities above 5 ft./sec. Use with caution.

PRESSURE LOSS PER 100' OF TYPE K COPPER PIPE (lbs./sq. in.)

C=130

GPM	3/4"	1"	1 1/4"	1 1/2"	2"	2 1/2"	3"	4"
1	0.232	0.057	0.019	0.008	0.002	0.001		
2	0.839	0.205	0.069	0.030	0.008	0.003	0.001	
3	1.777	0.435	0.146	0.063	0.016	0.006	0.002	
4	3.028	0.741	0.249	0.107	0.027	0.010	0.004	0.001
5	4.578	1.120	0.376	0.162	0.041	0.014	0.006	0.002
6	6.417	1.570	0.528	0.227	0.058	0.020	0.009	0.002
7	8.537	2.089	0.702	0.302	0.077	0.027	0.011	0.003
8	10.932	2.675	0.899	0.386	0.099	0.034	0.015	0.004
9	13.597	3.327	1.118	0.480	0.123	0.043	0.018	0.005
10	16.526	4.043	1.359	0.584	0.150	0.052	0.022	0.006
11	19.717	4.824	1.621	0.697	0.179	0.062	0.026	0.007
12		5.667	1.904	0.818	0.210	0.073	0.031	0.008
13		6.573	2.209	0.949	0.243	0.084	0.036	0.009
14		7.540	2.534	1.089	0.279	0.097	0.041	0.010
15		8.568	2.879	1.237	0.317	0.110	0.046	0.012
20		14.597	4.905	2.108	0.540	0.188	0.079	0.020
25			7.415	3.186	0.817	0.284	0.120	0.030
30			10.393	4.466	1.145	0.397	0.168	0.042
35			13.827	5.942	1.524	0.529	0.223	0.056
40			17.706	7.609	1.951	0.677	0.286	0.072
50				11.503	2.950	1.024	0.432	0.109
60				16.123	4.134	1.435	0.606	0.153
70					5.500	1.909	0.806	0.204
80					7.043	2.444	1.032	0.261
90					8.760	3.040	1.284	0.324
100					10.648	3.695	1.560	0.394
110					12.703	4.409	1.862	0.470
120					14.925	5.179	2.187	0.553
130					17.309	6.007	2.537	0.641
140					19.856	6.891	2.910	0.735
150						7.830	3.307	0.835
160						8.824	3.726	0.941
170						9.873	4.169	1.053
180						10.975	4.635	1.171
190						12.131	5.123	1.294
200						13.340	5.633	1.423
210						14.601	6.166	1.558
220						15.915	6.721	1.698
230						17.281	7.298	1.844
240						18.698	7.896	1.995

Shaded areas are velocities over 5 ft./sec. Use with caution.

VELOCITY OF FLOW FOR TYPE K COPPER PIPE
(ft./sec.)

GPM	3/4"	1"	1 1/4"	1 1/2"	2"	2 1/2"	3"	4"
1	0.735	0.412	0.263	0.186	0.106	0.069	0.048	0.027
2	1.471	0.825	0.527	0.372	0.213	0.138	0.097	0.055
3	2.206	1.237	0.790	0.558	0.319	0.207	0.145	0.082
4	2.942	1.649	1.053	0.744	0.425	0.275	0.193	0.110
5	3.677	2.061	1.317	0.930	0.532	0.344	0.241	0.137
6	4.412	2.474	1.580	1.117	0.638	0.413	0.290	0.165
7	5.148	2.886	1.843	1.303	0.744	0.482	0.338	0.192
8	5.883	3.298	2.107	1.489	0.851	0.551	0.386	0.219
9	6.619	3.710	2.370	1.675	0.957	0.620	0.435	0.247
10	7.354	4.123	2.633	1.861	1.064	0.688	0.483	0.274
11	8.089	4.535	2.897	2.047	1.170	0.757	0.531	0.302
12	8.825	4.947	3.160	2.233	1.276	0.826	0.580	0.329
13	9.560	5.360	3.423	2.419	1.383	0.895	0.628	0.357
14	10.296	5.772	3.687	2.605	1.489	0.964	0.676	0.384
15	11.031	6.184	3.950	2.791	1.595	1.033	0.724	0.412
20		8.246	5.267	3.722	2.127	1.377	0.966	0.549
25		10.307	6.583	4.652	2.659	1.721	1.207	0.686
30		12.368	7.900	5.583	3.191	2.065	1.449	0.823
35			9.216	6.513	3.722	2.409	1.690	0.960
40			10.533	7.444	4.254	2.754	1.932	1.097
50			13.166	9.305	5.318	3.442	2.415	1.372
60				11.165	6.381	4.130	2.898	1.646
70				13.026	7.445	4.819	3.381	1.921
80					8.509	5.507	3.864	2.195
90					9.572	6.196	4.347	2.469
100					10.636	6.884	4.830	2.744
110					11.699	7.572	5.313	3.018
120					12.763	8.261	5.796	3.292
130					13.826	8.949	6.279	3.567
140						9.637	6.762	3.841
150						10.326	7.245	4.116
160						11.014	7.728	4.390
170						11.703	8.211	4.664
180						12.391	8.694	4.939
190						13.079	9.177	5.213
200						13.768	9.660	5.487
210							10.143	5.762
220							10.626	6.036
230							11.109	6.310
240							11.592	6.585

Shaded areas are velocities over 5 ft./sec. Use with caution.

PRESSURE LOSS PER 100' OF TYPE L COPPER
PIPE (lbs./sq. in.)

C=130

GPM	3/4"	1"	1 1/4"	1 1/2"	2"	2 1/2"	3"	4"
1	0.180	0.049	0.018	0.008	0.002	0.001		
2	0.650	0.178	0.064	0.027	0.007	0.002	0.001	
3	1.378	0.376	0.135	0.058	0.015	0.005	0.002	
4	2.348	0.641	0.230	0.099	0.026	0.009	0.004	
5	3.549	0.969	0.348	0.150	0.039	0.014	0.006	0.001
6	4.975	1.359	0.488	0.210	0.055	0.019	0.008	0.002
7	6.619	1.808	0.649	0.279	0.073	0.025	0.011	0.003
8	8.476	2.315	0.832	0.357	0.093	0.032	0.014	0.003
9	10.542	2.879	1.034	0.444	0.116	0.040	0.017	0.004
10	12.813	3.499	1.257	0.540	0.140	0.049	0.021	0.005
11	15.287	4.175	1.500	0.644	0.168	0.058	0.025	0.006
12	17.960	4.905	1.762	0.757	0.197	0.069	0.029	0.007
13		5.689	2.044	0.878	0.228	0.080	0.033	0.008
14		6.525	2.345	1.007	0.262	0.091	0.038	0.010
15		7.415	2.664	1.144	0.298	0.104	0.044	0.011
20		12.632	4.539	1.949	0.507	0.177	0.074	0.019
25		19.097	6.861	2.947	0.766	0.267	0.112	0.028
30			9.617	4.130	1.074	0.374	0.158	0.040
35			12.795	5.495	1.429	0.498	0.210	0.053
40			16.385	7.037	1.830	0.638	0.268	0.068
50				10.638	2.766	0.964	0.406	0.103
60				14.910	3.877	1.352	0.569	0.144
70				19.837	5.158	1.798	0.757	0.192
80					6.606	2.303	0.969	0.246
90					8.216	2.864	1.205	0.305
100					9.986	3.482	1.465	0.371
110					11.914	4.154	1.748	0.443
120					13.997	4.880	2.053	0.520
130					16.234	5.660	2.381	0.603
140					18.622	6.492	2.732	0.692
150						7.377	3.104	0.787
160						8.314	3.498	0.886
170						9.302	3.914	0.992
180						10.340	4.351	1.103
190						11.429	4.809	1.219
200						12.568	5.288	1.340
210						13.757	5.789	1.467
220						14.995	6.309	1.599
230						16.281	6.851	1.736
240						17.617	7.413	1.878

Shaded areas are velocities over 5 ft./sec. Use with caution.

VELOCITY OF FLOW FOR TYPE L COPPER PIPE
(ft./sec.)

GPM	3/4"	1"	1 1/4"	1 1/2"	2"	2 1/2"	3"	4"
1	0.662	0.388	0.255	0.180	0.104	0.067	0.047	0.027
2	1.325	0.777	0.510	0.360	0.207	0.134	0.094	0.054
3	1.987	1.165	0.765	0.541	0.311	0.202	0.141	0.080
4	2.649	1.554	1.020	0.721	0.414	0.269	0.188	0.107
5	3.312	1.942	1.275	0.901	0.518	0.336	0.235	0.134
6	3.974	2.331	1.530	1.081	0.622	0.403	0.282	0.161
7	4.637	2.719	1.785	1.261	0.725	0.470	0.329	0.187
8	5.299	3.108	2.041	1.442	0.829	0.537	0.376	0.214
9	5.961	3.496	2.296	1.622	0.932	0.605	0.424	0.241
10	6.624	3.885	2.551	1.802	1.036	0.672	0.471	0.268
11	7.286	4.273	2.806	1.982	1.139	0.739	0.518	0.294
12	7.948	4.662	3.061	2.162	1.243	0.806	0.565	0.321
13	8.611	5.050	3.316	2.343	1.347	0.873	0.612	0.348
14	9.273	5.439	3.571	2.523	1.450	0.940	0.659	0.375
15	9.935	5.827	3.826	2.703	1.554	1.008	0.706	0.401
20	13.247	7.770	5.101	3.604	2.072	1.343	0.941	0.535
25		9.712	6.377	4.505	2.590	1.679	1.177	0.669
30		11.655	7.652	5.406	3.108	2.015	1.412	0.803
35		13.597	8.927	6.307	3.626	2.351	1.647	0.937
40			10.203	7.208	4.144	2.687	1.882	1.071
50			12.753	9.010	5.179	3.359	2.353	1.338
60				10.812	6.215	4.030	2.824	1.606
70				12.614	7.251	4.702	3.294	1.874
80					8.287	5.374	3.765	2.141
90					9.323	6.046	4.236	2.409
100					10.359	6.717	4.706	2.677
110					11.395	7.389	5.177	2.944
120					12.431	8.061	5.647	3.212
130					13.467	8.733	6.118	3.480
140						9.404	6.589	3.747
150						10.076	7.059	4.015
160						10.748	7.530	4.283
170						11.420	8.000	4.550
180						12.091	8.471	4.818
190						12.763	8.942	5.086
200						13.435	9.412	5.353
210							9.883	5.621
220							10.353	5.889
230							10.824	6.156
240							11.295	6.424

* Shaded areas are velocities over 5 ft./sec. Use with caution.

PRESSURE LOSS PER 100' OF TYPE M COPPER PIPE (lbs./sq. in.)

C=130

GPM	3/4"	1"	1 1/4"	1 1/2"	2"	2 1/2"	3"	4"
1	0.154	0.043	0.016	0.007	0.002	0.001		
2	0.555	0.154	0.058	0.026	0.007	0.002	0.001	
3	1.176	0.327	0.122	0.054	0.014	0.005	0.002	
4	2.004	0.557	0.209	0.092	0.024	0.008	0.004	
5	3.029	0.842	0.315	0.139	0.037	0.013	0.005	0.001
6	4.246	1.181	0.442	0.195	0.051	0.018	0.008	0.002
7	5.649	1.571	0.588	0.260	0.068	0.024	0.010	0.003
8	7.233	2.012	0.753	0.333	0.088	0.031	0.013	0.003
9	8.996	2.502	0.937	0.414	0.109	0.038	0.016	0.004
10	10.935	3.041	1.139	0.503	0.132	0.046	0.019	0.005
11	13.046	3.628	1.359	0.600	0.158	0.055	0.023	0.006
12	15.327	4.262	1.596	0.705	0.186	0.065	0.027	0.007
13	17.776	4.944	1.851	0.818	0.215	0.075	0.032	0.008
14		5.671	2.124	0.938	0.247	0.086	0.036	0.009
15		6.444	2.413	1.066	0.281	0.098	0.041	0.011
20		10.978	4.111	1.816	0.478	0.167	0.070	0.018
25		16.596	6.215	2.746	0.723	0.252	0.106	0.027
30			8.711	3.849	1.013	0.353	0.149	0.038
35			11.589	5.120	1.348	0.470	0.198	0.051
40			14.841	6.557	1.726	0.601	0.253	0.066
50				9.912	2.609	0.909	0.383	0.099
60				13.894	3.657	1.275	0.536	0.139
70				18.484	4.865	1.696	0.713	0.185
80					6.231	2.171	0.913	0.237
90					7.749	2.701	1.136	0.294
100					9.419	3.283	1.381	0.358
110					11.237	3.916	1.647	0.427
120					13.202	4.601	1.935	0.501
130					15.312	5.336	2.245	0.581
140					17.564	6.121	2.575	0.667
150					19.958	6.955	2.926	0.758
160						7.839	3.297	0.854
170						8.770	3.689	0.956
180						9.749	4.101	1.062
190						10.776	4.533	1.174
200						11.850	4.985	1.291
210						12.971	5.456	1.413
220						14.138	5.947	1.540
230						15.351	6.458	1.673
240						16.610	6.987	1.810

Shaded areas are velocities above 5 ft./sec. Use with caution.

VELOCITY OF FLOW FOR TYPE M COPPER PIPE
(ft./sec.)

GPM	3/4"	1"	1 1/4"	1 1/2"	2"	2 1/2"	3"	4"
1	0.621	0.367	0.245	0.175	0.101	0.066	0.046	0.026
2	1.241	0.733	0.490	0.350	0.202	0.131	0.092	0.053
3	1.862	1.100	0.735	0.525	0.303	0.197	0.138	0.079
4	2.482	1.467	0.980	0.700	0.405	0.262	0.184	0.105
5	3.103	1.834	1.224	0.875	0.506	0.328	0.230	0.132
6	3.723	2.200	1.469	1.050	0.607	0.393	0.276	0.158
7	4.344	2.567	1.714	1.225	0.708	0.459	0.322	0.185
8	4.965	2.934	1.959	1.400	0.809	0.525	0.367	0.211
9	5.585	3.300	2.204	1.575	0.910	0.590	0.413	0.237
10	6.206	3.667	2.449	1.750	1.011	0.656	0.459	0.264
11	6.826	4.034	2.694	1.926	1.112	0.721	0.505	0.290
12	7.447	4.401	2.939	2.101	1.214	0.787	0.551	0.316
13	8.067	4.767	3.184	2.276	1.315	0.852	0.597	0.343
14	8.688	5.134	3.429	2.451	1.416	0.918	0.643	0.369
15	9.309	5.501	3.673	2.626	1.517	0.984	0.689	0.395
20	12.411	7.334	4.898	3.501	2.023	1.311	0.919	0.527
25		9.168	6.122	4.376	2.528	1.639	1.148	0.659
30		11.001	7.347	5.251	3.034	1.967	1.378	0.791
35		12.835	8.571	6.127	3.540	2.295	1.608	0.923
40			9.796	7.002	4.045	2.623	1.837	1.054
50			12.245	8.752	5.056	3.278	2.297	1.318
60				10.503	6.068	3.934	2.756	1.582
70				12.253	7.079	4.590	3.215	1.845
80					8.090	5.245	3.675	2.109
90					9.102	5.901	4.134	2.372
100					10.113	6.557	4.593	2.636
110					11.124	7.212	5.052	2.900
120					12.135	7.868	5.512	3.163
130					13.147	8.524	5.971	3.427
140						9.180	6.430	3.690
150						9.835	6.890	3.954
160						10.491	7.349	4.218
170						11.147	7.808	4.481
180						11.802	8.268	4.745
190						12.458	8.727	5.008
200						13.114	9.186	5.272
210						13.769	9.646	5.536
220							10.105	5.799
230							10.564	6.063
240							11.024	6.326

Shaded areas are velocities above 5 ft./sec. Use with caution.

BRASS WATER SERVICE LINES - 1/2", 3/4", 1"

C=130, 1/2" ID=0.625, 3/4" ID=0.822, 1" ID=1.062

Flow	Velocity	PSI Loss	Velocity	PSI Loss	Velocity	PSI Loss
gpm	fps	psi / 100 ft.	fps	psi / 100 ft.	fps	psi / 100 ft.
	1/2"	1/2"	3/4"	3/4"	1"	1"
1	1.045	0.546	0.604	0.144	0.362	0.041
1.5	1.567	1.157	0.906	0.305	0.543	0.088
2	2.090	1.972	1.208	0.520	0.724	0.149
2.5	2.612	2.980	1.510	0.786	0.905	0.226
3	3.135	4.178	1.812	1.101	1.086	0.317
3.5	3.657	5.558	2.114	1.465	1.267	0.421
4	4.180	7.117	2.416	1.877	1.448	0.540
4.5	4.702	8.852	2.718	2.334	1.629	0.671
5	5.224	10.760	3.020	2.837	1.809	0.816
6	6.269	15.081	3.624	3.976	2.171	1.143
7	7.314	20.064	4.229	5.290	2.533	1.521
8	8.359	25.693	4.833	6.774	2.895	1.948
9	9.404	31.956	5.437	8.426	3.257	2.423
10	10.449	38.842	6.041	10.241	3.619	2.945
11	11.494	46.340	6.645	12.218	3.981	3.513
12	12.539	54.443	7.249	14.355	4.343	4.128
13	13.584	63.143	7.853	16.648	4.705	4.787
14	14.629	72.432	8.457	19.097	5.067	5.491
15	15.673	82.304	9.061	21.700	5.428	6.240
16	16.718	92.753	9.665	24.455	5.790	7.032
17	17.763	103.775	10.269	27.361	6.152	7.867
18	18.808	115.362	10.873	30.417	6.514	8.746
19	19.853	127.512	11.477	33.620	6.876	9.667
20			12.081	36.970	7.238	10.630
25			15.102	55.889	9.047	16.071
30			18.122	78.338	10.857	22.525
35					12.666	29.968
40					14.476	38.376
45					16.285	47.730
50					18.095	58.015

BRASS WATER SERVICE LINES - 1 ¹/₄", 1 ¹/₂", 2"

C=130, 1 1/4" ID=1.368, 1 1/2" ID=1.6, 2" ID=2.062

Flow	Velocity	PSI Loss	Velocity	PSI Loss	Velocity	PSI Loss
gpm	fps	psi / 100 ft.	fps	psi / 100 ft.	fps	psi / 100 ft.
	1 1/4"	1 1/4"	1 1/2"	1 1/2"	2"	2"
5	1.091	0.238	0.797	0.111	0.480	0.032
6	1.309	0.334	0.957	0.156	0.576	0.045
7	1.527	0.444	1.116	0.207	0.672	0.060
8	1.745	0.568	1.276	0.265	0.768	0.077
9	1.963	0.707	1.435	0.330	0.864	0.096
10	2.181	0.859	1.594	0.401	0.960	0.117
12	2.617	1.204	1.913	0.562	1.152	0.164
14	3.053	1.602	2.232	0.748	1.344	0.218
16	3.490	2.051	2.551	0.957	1.536	0.279
18	3.926	2.551	2.870	1.191	1.728	0.347
20	4.362	3.101	3.189	1.447	1.920	0.421
25	5.453	4.688	3.986	2.188	2.400	0.637
30	6.543	6.571	4.783	3.066	2.880	0.893
35	7.634	8.743	5.580	4.080	3.360	1.187
40	8.724	11.195	6.378	5.224	3.840	1.521
45	9.815	13.924	7.175	6.498	4.320	1.891
50	10.905	16.925	7.972	7.898	4.800	2.299
60	13.086	23.723	9.566	11.070	5.760	3.222
70	15.267	31.561	11.161	14.728	6.720	4.287
80	17.448	40.416	12.755	18.860	7.680	5.489
90	19.629	50.267	14.349	23.457	8.640	6.827
100			15.944	28.511	9.600	8.298
110			17.538	34.015	10.560	9.900
120			19.133	39.962	11.520	11.631
130					12.480	13.490
140					13.440	15.474
150					14.400	17.583

BRASS WATER SERVICE LINES - 2 ¹/₂", 3" 3 ¹/₂"

C=130, 2 1/2" ID=2.5, 3" ID=3.062, 3 1/2" ID=3.5

Flow	Velocity	PSI Loss	Velocity	PSI Loss	Velocity	PSI Loss
gpm	fps	psi / 100 ft.	fps	psi / 100 ft.	fps	psi / 100 ft.
	2 1/2"	2 1/2"	3"	3"	3 1/2"	3 1/2"
20	1.306	0.165	0.871	0.062	0.666	0.032
25	1.633	0.249	1.088	0.093	0.833	0.049
30	1.959	0.350	1.306	0.130	1.000	0.068
35	2.286	0.465	1.524	0.173	1.166	0.090
40	2.612	0.596	1.741	0.222	1.333	0.116
45	2.939	0.741	1.959	0.276	1.499	0.144
50	3.265	0.900	2.177	0.336	1.666	0.175
60	3.918	1.262	2.612	0.471	1.999	0.246
70	4.571	1.679	3.047	0.626	2.332	0.327
80	5.224	2.150	3.483	0.802	2.666	0.418
90	5.878	2.674	3.918	0.997	2.999	0.520
100	6.531	3.251	4.353	1.212	3.332	0.632
110	7.184	3.878	4.789	1.446	3.665	0.754
120	7.837	4.556	5.224	1.699	3.998	0.886
130	8.490	5.284	5.659	1.970	4.332	1.028
140	9.143	6.062	6.095	2.260	4.665	1.179
150	9.796	6.888	6.530	2.568	4.998	1.340
160	10.449	7.763	6.965	2.894	5.331	1.510
170	11.102	8.685	7.401	3.238	5.664	1.690
180	11.755	9.655	7.836	3.600	5.998	1.878
190	12.408	10.672	8.271	3.979	6.331	2.076
200	13.061	11.735	8.707	4.375	6.664	2.283
210	13.714	12.845	9.142	4.789	6.997	2.499
220	14.367	14.001	9.577	5.220	7.330	2.724
230	15.020	15.202	10.013	5.668	7.663	2.957
240	15.673	16.449	10.448	6.133	7.997	3.200
250	16.327	17.740	10.883	6.614	8.330	3.451
260	16.980	19.077	11.319	7.113	8.663	3.711
270	17.633	20.458	11.754	7.627	8.996	3.980
280	18.286	21.883	12.189	8.159	9.329	4.257
290	18.939	23.353	12.625	8.707	9.663	4.543
300	19.592	24.866	13.060	9.271	9.996	4.837
350			15.237	12.334	11.662	6.436
400			17.413	15.795	13.328	8.241
450			19.590	19.645	14.994	10.250
500			21.767	23.877	16.660	12.459

BRASS WATER SERVICE LINES - 4", 5", 6"

C=130, 4" ID=4.0, 5" ID=5.063, 6" ID=6.125

Flow	Velocity	PSI Loss	Velocity	PSI Loss	Velocity	PSI Loss
gpm	fps	psi / 100 ft.	fps	psi / 100 ft.	fps	psi / 100 ft.
	4"	4"	5"	5"	6"	6"
100	2.551	0.330	1.592	0.105	1.088	0.042
120	3.061	0.463	1.911	0.147	1.306	0.058
140	3.571	0.616	2.229	0.196	1.523	0.077
160	4.082	0.789	2.548	0.251	1.741	0.099
180	4.592	0.981	2.866	0.312	1.958	0.123
200	5.102	1.192	3.185	0.379	2.176	0.150
220	5.612	1.422	3.503	0.452	2.394	0.179
240	6.122	1.671	3.821	0.531	2.611	0.210
260	6.633	1.938	4.140	0.616	2.829	0.244
280	7.143	2.223	4.458	0.706	3.046	0.280
300	7.653	2.526	4.777	0.803	3.264	0.318
350	8.929	3.361	5.573	1.068	3.808	0.423
400	10.204	4.304	6.369	1.367	4.352	0.541
450	11.480	5.353	7.165	1.701	4.896	0.673
500	12.755	6.506	7.961	2.067	5.440	0.818
550	14.031	7.762	8.758	2.466	5.984	0.976
600	15.306	9.119	9.554	2.897	6.528	1.147
650	16.582	10.577	10.350	3.360	7.072	1.330
700	17.857	12.133	11.146	3.855	7.616	1.526
750	19.133	13.786	11.942	4.380	8.160	1.734
800	20.408	15.537	12.738	4.936	8.704	1.954
850	21.684	17.383	13.534	5.523	9.248	2.187
900	22.959	19.324	14.330	6.139	9.792	2.431
950	24.235	21.359	15.127	6.786	10.336	2.687
1000	25.510	23.487	15.923	7.462	10.880	2.955
1100	28.061	28.021	17.515	8.903	11.968	3.525
1200	30.612	32.921	19.107	10.459	13.056	4.141
1300	33.163	38.182	20.700	12.131	14.144	4.803
1400	35.714	43.799	22.292	13.915	15.232	5.510
1500	38.265	49.768	23.884	15.812	16.320	6.261
1600	40.816	56.087	25.476	17.819	17.408	7.055
1800	45.918	69.758	28.661	22.163	19.584	8.775
2000	51.020	84.789	31.845	26.938	21.760	10.666
2500			39.807	40.723	27.200	16.124
3000			47.768	57.080	32.639	22.600
3500			55.730	75.940	38.079	30.068
4000					43.519	38.504

POLYBUTYLENE (PB) WATER SERVICE
LINES - $^1/_2$", $^3/_4$", 1"

C=130, 1/2" ID=0.501, 3/4" ID=0.745, 1" ID=0.959

Flow	Velocity	PSI Loss	Velocity	PSI Loss	Velocity	PSI Loss
gpm	fps	psi / 100 ft.	fps	psi / 100 ft.	fps	psi / 100 ft.
	1/2"	1/2"	3/4"	3/4"	1"	1"
1	1.626	1.602	0.735	0.232	0.444	0.068
1.5	2.439	3.394	1.103	0.492	0.666	0.144
2	3.252	5.782	1.471	0.839	0.888	0.246
2.5	4.065	8.741	1.838	1.268	1.110	0.371
3	4.878	12.252	2.206	1.777	1.331	0.520
3.5	5.691	16.301	2.574	2.365	1.553	0.692
4	6.505	20.874	2.942	3.028	1.775	0.886
4.5	7.318	25.962	3.309	3.766	1.997	1.102
5	8.131	31.556	3.677	4.578	2.219	1.340
6	9.757	44.231	4.412	6.417	2.663	1.878
7	11.383	58.846	5.148	8.537	3.107	2.499
8	13.009	75.356	5.883	10.932	3.550	3.200
9	14.635	93.724	6.619	13.597	3.994	3.980
10	16.261	113.918	7.354	16.526	4.438	4.837
11	17.888	135.910	8.089	19.717	4.882	5.771
12	19.514	159.675	8.825	23.164	5.326	6.780
13			9.560	26.866	5.770	7.864
14			10.296	30.818	6.213	9.021
15			11.031	35.019	6.657	10.250
16			11.766	39.465	7.101	11.552
17			12.502	44.154	7.545	12.924
18			13.237	49.084	7.989	14.367
19			13.973	54.254	8.432	15.881
20			14.708	59.660	8.876	17.463
25			18.385	90.191	11.095	26.400
30			22.062	126.417	13.314	37.004
35			25.739	168.187	15.533	49.230
40					17.752	63.042
45					19.971	78.409
50					22.190	95.303

POLYBUTYLENE (PB) WATER SERVICE
LINE - 1^1/$_4$", 1^1/$_2$", 2"

C=130, 1 1/4" ID=1.171, 1 1/2" ID=1.385, 2" ID=1.811

Flow	Velocity	PSI Loss	Velocity	PSI Loss	Velocity	PSI Loss
gpm	fps	psi / 100 ft.	fps	psi / 100 ft.	fps	psi / 100 ft.
	1 1/4"	1 1/4"	1 1/2"	1 1/2"	2"	2"
2	0.595	0.093	0.426	0.041	0.249	0.011
4	1.191	0.335	0.851	0.148	0.498	0.040
6	1.786	0.711	1.277	0.314	0.747	0.085
8	2.381	1.211	1.702	0.535	0.996	0.145
10	2.977	1.831	2.128	0.809	1.245	0.219
12	3.572	2.566	2.553	1.134	1.493	0.308
14	4.167	3.414	2.979	1.509	1.742	0.409
16	4.763	4.371	3.405	1.932	1.991	0.524
18	5.358	5.437	3.830	2.403	2.240	0.652
20	5.953	6.608	4.256	2.920	2.489	0.792
25	7.441	9.990	5.320	4.415	3.111	1.197
30	8.930	14.003	6.383	6.188	3.734	1.678
35	10.418	18.630	7.447	8.233	4.356	2.233
40	11.906	23.856	8.511	10.543	4.978	2.859
45	13.395	29.672	9.575	13.112	5.600	3.556
50	14.883	36.065	10.639	15.938	6.223	4.323
60	17.860	50.551	12.767	22.339	7.467	6.059
70	20.836	67.253	14.895	29.720	8.712	8.061
80	23.813	86.122	17.023	38.059	9.956	10.322
90	26.789	107.114	19.150	47.336	11.201	12.838
100	29.766	130.194	21.278	57.535	12.445	15.605
110	32.743	155.328	23.406	68.642	13.690	18.617
120	35.719	182.488	25.534	80.644	14.934	21.872
130	38.696	211.648	27.662	93.531	16.179	25.367
140	41.672	242.784	29.789	107.290	17.423	29.099
150	44.649	275.874	31.917	121.913	18.668	33.065
160	47.626	310.900	34.045	137.392	19.912	37.263
170	50.602	347.842	36.173	153.717	21.157	41.691
180	53.579	386.683	38.301	170.881	22.401	46.346
190	56.555	427.408	40.429	188.878	23.646	51.228
200	59.532	470.000	42.556	207.701	24.890	56.333

CAST IRON WATER LINES - 3", 3¹/₂", 4"

C=100, 3" ID=3.0, 3 1/2" ID=3.5, 4" ID=4.0

Flow	Velocity	PSI Loss	Velocity	PSI Loss	Velocity	PSI Loss
gpm	fps	psi / 100 ft.	fps	psi / 100 ft.	fps	psi / 100 ft.
	3"	3"	3 1/2"	3 1/2"	4"	4"
10	0.454	0.031	0.333	0.014	0.255	0.008
15	0.680	0.065	0.500	0.031	0.383	0.016
20	0.907	0.110	0.666	0.052	0.510	0.027
25	1.134	0.167	0.833	0.079	0.638	0.041
30	1.361	0.234	1.000	0.111	0.765	0.058
35	1.587	0.311	1.166	0.147	0.893	0.077
40	1.814	0.399	1.333	0.188	1.020	0.098
45	2.041	0.496	1.499	0.234	1.148	0.122
50	2.268	0.603	1.666	0.285	1.276	0.149
60	2.721	0.845	1.999	0.399	1.531	0.208
70	3.175	1.124	2.332	0.531	1.786	0.277
80	3.628	1.440	2.666	0.680	2.041	0.355
90	4.082	1.791	2.999	0.846	2.296	0.442
100	4.535	2.176	3.332	1.028	2.551	0.537
110	4.989	2.597	3.665	1.227	2.806	0.640
120	5.442	3.051	3.998	1.441	3.061	0.752
130	5.896	3.538	4.332	1.671	3.316	0.873
140	6.349	4.059	4.665	1.917	3.571	1.001
150	6.803	4.612	4.998	2.178	3.827	1.138
160	7.256	5.197	5.331	2.455	4.082	1.282
170	7.710	5.815	5.664	2.747	4.337	1.434
180	8.163	6.464	5.998	3.053	4.592	1.594
190	8.617	7.145	6.331	3.375	4.847	1.762
200	9.070	7.857	6.664	3.711	5.102	1.938
250	11.338	11.878	8.330	5.610	6.378	2.930
300	13.605	16.648	9.996	7.864	7.653	4.107
350	15.873	22.149	11.662	10.462	8.929	5.463
400	18.141	28.363	13.328	13.398	10.204	6.996
450	20.408	35.277	14.994	16.663	11.480	8.702
500	22.676	42.878	16.660	20.254	12.755	10.577
550	24.943	51.156	18.326	24.164	14.031	12.618
600	27.211	60.101	19.992	28.389	15.306	14.825
650	29.478	69.704	21.658	32.925	16.582	17.194
700	31.746	79.959	23.324	37.769	17.857	19.723
750	34.014	90.857	24.990	42.917	19.133	22.411

CAST IRON WATER LINE - 6", 8", 10"

C=100, 6" ID=6.0, 8" ID=8.0, 10" ID=10.0

Flow	Velocity	PSI Loss	Velocity	PSI Loss	Velocity	PSI Loss
gpm	fps	psi / 100 ft.	fps	psi / 100 ft.	fps	psi / 100 ft.
	6"	6"	8"	8"	10"	10"
50	0.567	0.021	0.319	0.005	0.204	0.002
100	1.134	0.075	0.638	0.018	0.408	0.006
150	1.701	0.158	0.957	0.039	0.612	0.013
200	2.268	0.270	1.276	0.066	0.816	0.022
250	2.834	0.407	1.594	0.101	1.020	0.034
300	3.401	0.571	1.913	0.141	1.224	0.048
350	3.968	0.760	2.232	0.187	1.429	0.063
400	4.535	0.973	2.551	0.240	1.633	0.081
450	5.102	1.210	2.870	0.298	1.837	0.101
500	5.669	1.471	3.189	0.363	2.041	0.123
550	6.236	1.755	3.508	0.433	2.245	0.146
600	6.803	2.062	3.827	0.509	2.449	0.172
650	7.370	2.391	4.145	0.590	2.653	0.199
700	7.937	2.743	4.464	0.677	2.857	0.228
750	8.503	3.117	4.783	0.769	3.061	0.260
800	9.070	3.512	5.102	0.866	3.265	0.293
850	9.637	3.930	5.421	0.969	3.469	0.327
900	10.204	4.369	5.740	1.078	3.673	0.364
950	10.771	4.829	6.059	1.191	3.878	0.402
1000	11.338	5.310	6.378	1.310	4.082	0.442
1100	12.472	6.335	7.015	1.563	4.490	0.528
1200	13.605	7.443	7.653	1.836	4.898	0.620
1300	14.739	8.632	8.291	2.129	5.306	0.719
1400	15.873	9.902	8.929	2.442	5.714	0.825
1500	17.007	11.251	9.566	2.775	6.122	0.937
1600	18.141	12.680	10.204	3.128	6.531	1.056
1700	19.274	14.186	10.842	3.499	6.939	1.182
1800	20.408	15.771	11.480	3.890	7.347	1.314
1900	21.542	17.431	12.117	4.300	7.755	1.452
2000	22.676	19.169	12.755	4.728	8.163	1.597
2500	28.345	28.978	15.944	7.148	10.204	2.414
3000	34.014	40.617	19.133	10.019	12.245	3.383
3500	39.683	54.038	22.321	13.329	14.286	4.501
4000	45.351	69.199	25.510	17.069	16.327	5.764
4500	51.020	86.066	28.699	21.230	18.367	7.168
5000	56.689	104.610	31.888	25.804	20.408	8.713

STEEL WATER LINES - $^3/_4$", 1", $1^1/_4$"

C=100, 3/4" ID=0.824, 1" ID=1.049, 1 1/4" ID=1.380

Flow	Velocity	PSI Loss	Velocity	PSI Loss	Velocity	PSI Loss
gpm	fps	psi / 100 ft.	fps	psi / 100 ft.	fps	psi / 100 ft.
	3/4"	3/4"	1"	1"	1 1/4"	1 1/4"
1	0.601	0.231	0.371	0.071	0.214	0.019
1.5	0.902	0.490	0.556	0.151	0.321	0.040
2	1.202	0.835	0.742	0.258	0.429	0.068
2.5	1.503	1.262	0.927	0.390	0.536	0.103
3	1.803	1.770	1.113	0.547	0.643	0.144
3.5	2.104	2.354	1.298	0.727	0.750	0.192
4	2.405	3.015	1.484	0.931	0.857	0.245
4.5	2.705	3.750	1.669	1.158	0.964	0.305
5	3.006	4.557	1.855	1.408	1.072	0.371
6	3.607	6.388	2.226	1.973	1.286	0.520
7	4.208	8.499	2.596	2.626	1.500	0.691
8	4.809	10.883	2.967	3.362	1.715	0.885
9	5.410	13.536	3.338	4.182	1.929	1.101
10	6.011	16.453	3.709	5.083	2.143	1.338
12	7.214	23.061	4.451	7.124	2.572	1.876
14	8.416	30.681	5.193	9.478	3.001	2.496
16	9.618	39.288	5.935	12.137	3.429	3.196
18	10.821	48.865	6.677	15.096	3.858	3.975
20	12.023	59.394	7.418	18.349	4.287	4.832
30	18.034	125.853	11.128	38.880	6.430	10.238
40	24.046	214.412	14.837	66.238	8.573	17.443
50			18.546	100.135	10.716	26.369
60			22.255	140.356	12.860	36.960
70			25.965	186.731	15.003	49.172
80					17.146	62.968
90					19.289	78.316
100					21.433	95.191

STEEL WATER LINES - 1¹/₂", 2", 2¹/₂"

C=100, 1 1/2" ID=1.610, 2" ID=2.067, 2 1/2" ID=2.469

Flow	Velocity	PSI Loss	Velocity	PSI Loss	Velocity	PSI Loss
gpm	fps	psi / 100 ft.	fps	psi / 100 ft.	fps	psi / 100 ft.
	1 1/2"	1 1/2"	2"	2"	2 1/2"	2 1/2"
5	0.787	0.175	0.478	0.052	0.335	0.022
6	0.945	0.245	0.573	0.073	0.402	0.031
7	1.102	0.327	0.669	0.097	0.469	0.041
8	1.260	0.418	0.764	0.124	0.536	0.052
9	1.417	0.520	0.860	0.154	0.603	0.065
10	1.575	0.632	0.955	0.187	0.670	0.079
12	1.890	0.886	1.146	0.263	0.803	0.111
14	2.205	1.179	1.337	0.350	0.937	0.147
16	2.519	1.510	1.529	0.448	1.071	0.189
18	2.834	1.878	1.720	0.557	1.205	0.234
20	3.149	2.282	1.911	0.677	1.339	0.285
25	3.937	3.450	2.388	1.023	1.674	0.431
30	4.724	4.836	2.866	1.434	2.009	0.604
35	5.511	6.434	3.344	1.908	2.343	0.803
40	6.299	8.239	3.821	2.443	2.678	1.029
45	7.086	10.247	4.299	3.038	3.013	1.280
50	7.873	12.455	4.777	3.693	3.348	1.555
60	9.448	17.458	5.732	5.176	4.017	2.180
70	11.023	23.227	6.687	6.887	4.687	2.901
80	12.597	29.743	7.643	8.819	5.357	3.714
90	14.172	36.993	8.598	10.968	6.026	4.620
100	15.746	44.964	9.553	13.332	6.696	5.615
200	31.493	162.321	19.107	48.128	13.391	20.271
300			28.660	101.980	20.087	42.953
400			38.213	173.741	26.783	73.178
500					33.478	110.626

STEEL WATER LINES - 3", 3 ½", 4"

C=100, 3" ID=3.068, 3 1/2" ID=3.548, 4" ID=4.026

Flow	Velocity	PSI Loss	Velocity	PSI Loss	Velocity	PSI Loss
gpm	fps	psi / 100 ft.	fps	psi / 100 ft.	fps	psi / 100 ft.
	3"	3"	3 1/2"	3 1/2"	4"	4"
20	0.867	0.099	0.648	0.049	0.504	0.026
25	1.084	0.150	0.811	0.074	0.630	0.040
30	1.301	0.210	0.973	0.103	0.755	0.056
35	1.518	0.279	1.135	0.138	0.881	0.074
40	1.735	0.358	1.297	0.176	1.007	0.095
45	1.951	0.445	1.459	0.219	1.133	0.119
50	2.168	0.541	1.621	0.267	1.259	0.144
60	2.602	0.758	1.945	0.374	1.511	0.202
70	3.035	1.008	2.270	0.497	1.763	0.269
80	3.469	1.291	2.594	0.636	2.015	0.344
90	3.903	1.606	2.918	0.792	2.266	0.428
100	4.336	1.952	3.242	0.962	2.518	0.520
150	6.505	4.135	4.864	2.039	3.777	1.102
200	8.673	7.045	6.485	3.473	5.036	1.878
250	10.841	10.650	8.106	5.251	6.295	2.839
300	13.009	14.928	9.727	7.360	7.555	3.979
350	15.177	19.861	11.348	9.791	8.814	5.294
400	17.345	25.433	12.970	12.539	10.073	6.779
450	19.514	31.632	14.591	15.595	11.332	8.432
500	21.682	38.448	16.212	18.955	12.591	10.248
550	23.850	45.871	17.833	22.614	13.850	12.227
600	26.018	53.891	19.454	26.569	15.109	14.365
650	28.186	62.503	21.076	30.814	16.368	16.660
700	30.354	71.697	22.697	35.347	17.627	19.111
750	32.523	81.470	24.318	40.165	18.886	21.716
800	34.691	91.813	25.939	45.264	20.145	24.473
850	36.859	102.723	27.560	50.643	21.405	27.381
900	39.027	114.193	29.182	56.297	22.664	30.438
950			30.803	62.227	23.923	33.644
1000			32.424	68.428	25.182	36.997
1100			35.666	81.638	27.700	44.139
1200			38.909	95.913	30.218	51.857
1300					32.736	60.143
1400					35.254	68.991
1500					37.773	78.394

STEEL WATER LINES - 5", 6", 8"

C=100, 5" ID=5.047, 6" ID=6.065, 8" ID=7.981

Flow	Velocity	PSI Loss	Velocity	PSI Loss	Velocity	PSI Loss
gpm	fps	psi / 100 ft.	fps	psi / 100 ft.	fps	psi / 100 ft.
	5"	5"	6"	6"	8"	8"
50	0.801	0.048	0.555	0.020	0.320	0.005
60	0.961	0.067	0.666	0.028	0.384	0.007
70	1.122	0.089	0.777	0.037	0.449	0.010
80	1.282	0.115	0.888	0.047	0.513	0.012
90	1.442	0.143	0.999	0.058	0.577	0.015
100	1.602	0.173	1.110	0.071	0.641	0.019
150	2.404	0.367	1.664	0.150	0.961	0.039
200	3.205	0.625	2.219	0.256	1.282	0.067
250	4.006	0.945	2.774	0.387	1.602	0.102
300	4.807	1.325	3.329	0.542	1.922	0.143
350	5.608	1.763	3.884	0.721	2.243	0.190
400	6.410	2.257	4.438	0.923	2.563	0.243
450	7.211	2.807	4.993	1.148	2.884	0.302
500	8.012	3.412	5.548	1.396	3.204	0.367
550	8.813	4.071	6.103	1.665	3.524	0.438
600	9.614	4.783	6.658	1.956	3.845	0.514
650	10.416	5.547	7.212	2.269	4.165	0.597
700	11.217	6.363	7.767	2.603	4.486	0.684
750	12.018	7.231	8.322	2.958	4.806	0.778
800	12.819	8.149	8.877	3.333	5.126	0.876
850	13.620	9.117	9.432	3.729	5.447	0.981
900	14.421	10.135	9.987	4.145	5.767	1.090
950	15.223	11.202	10.541	4.582	6.088	1.205
1000	16.024	12.319	11.096	5.039	6.408	1.325
1100	17.626	14.697	12.206	6.011	7.049	1.581
1200	19.229	17.267	13.315	7.062	7.690	1.857
1300	20.831	20.026	14.425	8.191	8.330	2.154
1400	22.433	22.972	15.535	9.396	8.971	2.471
1500	24.036	26.103	16.644	10.677	9.612	2.808
1600	25.638	29.417	17.754	12.032	10.253	3.164
1800	28.843	36.588	19.973	14.965	11.534	3.935
2000	32.048	44.471	22.192	18.190	12.816	4.783
2500			27.740	27.498	16.020	7.231
3000			33.288	38.543	19.224	10.136
3500			38.837	51.278	22.428	13.484
4000					25.632	17.268
4500					28.836	21.477
5000					32.040	26.104

STEEL WATER LINES - 10", 12"

C=100, 10" ID=10.02, 12" ID=12.0

Flow	Velocity	PSI Loss	Velocity	PSI Loss
gpm	fps	psi / 100 ft.	fps	psi / 100 ft.
	10"	10"	12"	12"
150	0.610	0.013	0.425	0.005
200	0.813	0.022	0.567	0.009
250	1.016	0.034	0.709	0.014
300	1.220	0.047	0.850	0.020
350	1.423	0.063	0.992	0.026
400	1.626	0.080	1.134	0.033
450	1.829	0.100	1.276	0.042
500	2.033	0.121	1.417	0.050
600	2.439	0.170	1.701	0.071
700	2.846	0.226	1.984	0.094
800	3.252	0.290	2.268	0.120
900	3.659	0.360	2.551	0.150
1000	4.065	0.438	2.834	0.182
1200	4.878	0.614	3.401	0.255
1400	5.691	0.817	3.968	0.340
1600	6.505	1.046	4.535	0.435
1800	7.318	1.301	5.102	0.541
2000	8.131	1.581	5.669	0.658
2500	10.163	2.390	7.086	0.994
3000	12.196	3.350	8.503	1.393
3500	14.229	4.457	9.921	1.854
4000	16.261	5.708	11.338	2.374
4500	18.294	7.099	12.755	2.952
5000	20.327	8.629	14.172	3.589
5500	22.359	10.295	15.590	4.281
6000	24.392	12.095	17.007	5.030
6500	26.425	14.027	18.424	5.834
7000	28.457	16.091	19.841	6.692
7500	30.490	18.284	21.259	7.604
8000	32.523	20.605	22.676	8.569
8500	34.556	23.054	24.093	9.587
9000	36.588	25.628	25.510	10.658
9500	38.621	28.327	26.927	11.781
10000	40.654	31.150	28.345	12.955
11000			31.179	15.455
12000			34.014	18.158
13000			36.848	21.059
14000			39.683	24.157

PRESSURE LOSS IN PIPE FITTINGS AND VALVES

(approximate equivalent in feet of pipe)

Plastic Fittings:

Size	45° Elbow	90° Elbow	Tee Side Outlet	Run of Straight Tee
1/2"	1.5	3.0	5.9	1.9
3/4"	2.0	3.8	7.4	2.5
1"	2.7	4.9	10.1	3.4
1 1/4"	3.2	6.1	12.1	4.0
1 1/2"	3.5	6.4	13.0	4.3
2"	4.0	7.4	14.7	5.0
2 1/2"	5.0	9.4	18.7	6.2
3"	5.7	10.8	21.4	7.1
4"	6.8	12.8	25.1	8.4
6"	9.1	17.0	33.9	11.3
8"	11.4	21.6	43.1	14.4

Steel Fittings:

Size	45° Elbow	90° Elbow	Tee Side Outlet	Run of Straight Tee
1/2"	0.8	1.6	3.1	1.0
3/4"	1.1	2.1	4.1	1.4
1"	1.4	2.6	5.3	1.8
1 1/4"	1.8	3.5	6.9	2.3
1 1/2"	2.2	4.0	8.1	2.7
2"	2.8	5.2	10.3	3.5
2 1/2"	3.3	6.2	12.3	4.1
3"	4.1	7.7	15.3	5.1
4"	5.4	10.2	20.1	6.7
6"	8.1	15.2	30.3	10.1
8"	10.6	20.0	39.9	13.3

Valves:

Size	Corporation Stop	Curb Stop	Globe Valve	Gate Valve
1/2"	6.1	3.1	17.6	0.4
3/4"	5.9	4	23.3	0.6
1"	6.7	3.8	29.7	0.7
1 1/4"	7.5	3.6	39.1	0.9
1 1/2"	7.7	4.4	45.6	1.1
2"	8.4	4.8	58.6	1.4

AVERAGE METER FLOWS AND PRESSURE LOSSES

AWWA Standards M22 "Sizing Water Service Lines and Meters"
Displacement type meters

Size	Maximum Capacity		* 70% of Maximum Capacity		** 30% of Maximum Capacity	
in.	gpm	psi	gpm	psi	gpm	psi
5/8	20	10.4	14	5.1	6	1.0
3/4	30	10.6	21	5.3	9	1.1
1	50	9.3	35	5.0	17	1.1
1 1/2	100	11.3	70	7.1	30	0.9
2	160	10.4	112	5.6	38	0.5
3	300	13.1	210	6.7	90	1.1
4***	500	9.6	350	5.4	250	3.5

Recommended design flow for irrigation systems.

**Recommended for continuous flow based on AWWA Manual M22.*

***Compound type meter with a continous flow rate of 50 percent.

AVERAGE PRESSURE LOSS THROUGH STANDARD WATER METERS (LBS./SQ. IN.)

GPM	5/8"	3/4"	1"	1 1/2"	2"	3"	4"
5	0.7	0.5					
6	1.0	0.6					
7	1.4	0.7					
8	1.9	0.9					
9	2.4	1.1					
10	2.9	1.3					
11	3.5	1.6					
12	3.9	1.9	0.1				
13	4.5	2.3	0.3				
14	5.1	2.6	0.5				
15	5.5	3.0	0.7				
16	6.1	3.4	1.0				
17	7.0	3.7	1.3				
18	8.1	4.1	1.6				
19	9.2	4.4	1.9				
20	10.4	4.9	2.2	0.4			
25		7.5	2.9	0.7			
30		10.6	3.7	0.9	0.2		
35			5.0	1.5	0.4		
40			6.3	2.3	0.6		
50			9.3	3.8	1.1		
60				5.4	1.7		
70				7.1	2.3	0.5	
80				8.6	3.0	0.8	0.6
90				10.0	3.9	1.1	0.8
100				11.3	5.0	1.5	1.0
120					6.1	2.3	1.4
140					8.7	3.1	1.8
160					10.4	4.2	2.2
180						5.1	2.5
200						6.2	2.8
220						7.1	3.0
240						8.3	3.3
260						9.9	3.6
280						11.5	3.9
300						13.1	4.3
320							4.7
340							5.2
360							5.6
380							6.0
400							6.3
450							8.0
500							9.6

STRAINER SIZES

Mesh Size	Microns	Inches
4	5205	0.2030
8	2487	0.0970
10	1923	0.0750
14	1307	0.0510
18	1000	0.0394
20	840	0.0331
25	710	0.0280
30	590	0.0232
35	500	0.0197
40	420	0.0165
45	350	0.0138
50	297	0.0117
60	250	0.0098
70	210	0.0083
80	177	0.0070
100	149	0.0059
120	125	0.0049
140	105	0.0041
170	88	0.0035
200	74	0.0029
230	62	0.0024
270	53	0.0021
325	44	0.0017
400	37	0.0015
550	25	0.0009
800	15	0.0006
1250	10	0.0004

AVERAGE APPLICATION RATES (in./hr.)

Equilateral Triangular Spacing

Spacing	Discharge Rates per Sprinkler (gpm)														
feet	1	2	3	4	5	6	7	8	9	10	12	15	20	25	30
10	1.11	2.22	3.33	4.45	5.56										
11	0.92	1.84	2.76	3.67	4.59										
12	0.77	1.54	2.32	3.09	3.86	4.63									
13	0.66	1.32	1.97	2.63	3.29	3.95									
14	0.57	1.13	1.70	2.27	2.84	3.40	3.97								
15	0.49	0.99	1.48	1.98	2.47	2.96	3.46								
16	0.43	0.87	1.30	1.74	2.17	2.60	3.04	3.47							
17	0.38	0.77	1.15	1.54	1.92	2.31	2.69	3.08							
18	0.34	0.69	1.03	1.37	1.72	2.06	2.40	2.74	3.09						
19	0.31	0.62	0.92	1.23	1.54	1.85	2.16	2.46	2.77						
20	0.28	0.56	0.83	1.11	1.39	1.67	1.95	2.22	2.50	2.78					
21	0.25	0.50	0.76	1.01	1.26	1.51	1.76	2.02	2.27	2.52					
22	0.23	0.46	0.69	0.92	1.15	1.38	1.61	1.84	2.07	2.30	2.76				
23	0.21	0.42	0.63	0.84	1.05	1.26	1.47	1.68	1.89	2.10	2.52				
24	0.19	0.39	0.58	0.77	0.96	1.16	1.35	1.54	1.74	1.93	2.32				
25	0.18	0.36	0.53	0.71	0.89	1.07	1.24	1.42	1.60	1.78	2.13	2.67	3.56		
30	0.12	0.25	0.37	0.49	0.62	0.74	0.86	0.99	1.11	1.23	1.48	1.85	2.47		
35	0.09	0.18	0.27	0.36	0.45	0.54	0.64	0.73	0.82	0.91	1.09	1.36	1.81	2.27	2.72
40	0.07	0.14	0.21	0.28	0.35	0.42	0.49	0.56	0.63	0.69	0.83	1.04	1.39	1.74	2.08
45	0.05	0.11	0.16	0.22	0.27	0.33	0.38	0.44	0.49	0.55	0.66	0.82	1.10	1.37	1.65
50	0.04	0.09	0.13	0.18	0.22	0.27	0.31	0.36	0.40	0.44	0.53	0.67	0.89	1.11	1.33
55	0.04	0.07	0.11	0.15	0.18	0.22	0.26	0.29	0.33	0.37	0.44	0.55	0.73	0.92	1.10
60	0.03	0.06	0.09	0.12	0.15	0.19	0.22	0.25	0.28	0.31	0.37	0.46	0.62	0.77	0.93
65	0.03	0.05	0.08	0.11	0.13	0.16	0.18	0.21	0.24	0.26	0.32	0.39	0.53	0.66	0.79
70	0.02	0.05	0.07	0.09	0.11	0.14	0.16	0.18	0.20	0.23	0.27	0.34	0.45	0.57	0.68
75	0.02	0.04	0.06	0.08	0.10	0.12	0.14	0.16	0.18	0.20	0.24	0.30	0.40	0.49	0.59

AVERAGE APPLICATIONS (in./hr.)

Square and Rectangular Spacing

Spacing feet	Discharge Rates per Sprinkler (gpm)														
	1	2	3	4	5	6	7	8	9	10	12	15	20	25	30
10 x 10	0.96	1.93	2.89	3.85	4.81	5.78	6.74	7.70							
15 x 15	0.43	0.86	1.28	1.71	2.14	2.57	2.99	3.42	3.85						
20 x 20	0.24	0.48	0.72	0.96	1.20	1.44	1.68	1.93	2.17	2.41					
20 x 30	0.16	0.32	0.48	0.64	0.80	0.96	1.12	1.28	1.44	1.60	1.93				
20 x 40	0.12	0.24	0.36	0.48	0.60	0.72	0.84	0.96	1.08	1.20	1.44	1.80			
20 x 50	0.10	0.19	0.29	0.39	0.48	0.58	0.67	0.77	0.87	0.96	1.16	1.44	1.93		
25 x 25	0.15	0.31	0.46	0.62	0.77	0.92	1.08	1.23	1.39	1.54	1.85	2.31	3.08	3.85	
30 x 30	0.11	0.21	0.32	0.43	0.53	0.64	0.75	0.86	0.96	1.07	1.28	1.60	2.14	2.67	3.21
30 x 40	0.08	0.16	0.24	0.32	0.40	0.48	0.56	0.64	0.72	0.80	0.96	1.20	1.60	2.01	2.41
30 x 50	0.06	0.13	0.19	0.26	0.32	0.39	0.45	0.51	0.58	0.64	0.77	0.96	1.28	1.60	1.93
40 x 40	0.06	0.12	0.18	0.24	0.30	0.36	0.42	0.48	0.54	0.60	0.72	0.90	1.20	1.50	1.80
40 x 50		0.10	0.14	0.19	0.24	0.29	0.34	0.39	0.43	0.48	0.58	0.72	0.96	1.20	1.44
40 x 60			0.12	0.16	0.20	0.24	0.28	0.32	0.36	0.40	0.48	0.60	0.80	1.00	1.20
50 x 50			0.12	0.15	0.19	0.23	0.27	0.31	0.35	0.39	0.46	0.58	0.77	0.96	1.16
50 x 60			0.10	0.13	0.16	0.19	0.22	0.26	0.29	0.32	0.39	0.48	0.64	0.80	0.96
50 x 70				0.11	0.14	0.17	0.19	0.22	0.25	0.28	0.33	0.41	0.55	0.69	0.83
50 x 80				0.10	0.12	0.14	0.17	0.19	0.22	0.24	0.29	0.36	0.48	0.60	0.72
60 x 60				0.11	0.13	0.16	0.19	0.21	0.24	0.27	0.32	0.40	0.53	0.67	0.80
60 x 70					0.11	0.14	0.16	0.18	0.21	0.23	0.28	0.34	0.46	0.57	0.69
60 x 80						0.12	0.14	0.16	0.18	0.20	0.24	0.30	0.40	0.50	0.60
70 x 70						0.12	0.14	0.16	0.18	0.20	0.24	0.29	0.39	0.49	0.59
70 x 80						0.10	0.12	0.14	0.15	0.17	0.21	0.26	0.34	0.43	0.52
70 x 90								0.12	0.14	0.15	0.18	0.23	0.31	0.38	0.46
80 x 80								0.12	0.14	0.15	0.18	0.23	0.30	0.38	0.45

CONVERSION FROM PRESSURE (psi) TO FEET OF HEAD

Pressure psi	Head feet	Pressure psi	Head feet	Pressure psi	Head feet
1	2.310	78	180.180	280	646.800
2	4.620	80	184.800	285	658.350
3	6.930	90	207.900	290	669.900
4	9.240	92	212.520	295	681.450
5	11.550	94	217.140	300	693.000
6	13.860	96	221.760	305	704.550
7	16.170	98	226.380	310	716.100
8	18.480	100	231.000	315	727.650
9	20.790	105	242.550	320	739.200
10	23.100	110	254.100	325	750.750
12	27.720	115	265.650	330	762.300
14	32.340	120	277.200	335	773.850
16	36.960	125	288.750	340	785.400
18	41.580	130	300.300	345	796.950
20	46.200	135	311.850	350	808.500
22	50.820	140	323.400	355	820.050
24	55.440	145	334.950	360	831.600
26	60.060	150	346.500	365	843.150
28	64.680	155	358.050	370	854.700
30	69.300	160	369.600	375	866.250
32	73.920	165	381.150	380	877.800
34	78.540	170	392.700	385	889.350
36	83.160	175	404.250	390	900.900
38	87.780	180	415.800	395	912.450
40	92.400	185	427.350	400	924.000
42	97.020	190	438.900	405	935.550
44	101.640	195	450.450	410	947.100
46	106.260	200	462.000	415	958.650
48	110.880	205	473.550	420	970.200
50	115.500	210	485.100	425	981.750
52	120.120	215	496.650	430	993.300
54	124.740	220	508.200	435	1004.850
56	129.360	225	519.750	440	1016.400
58	133.980	230	531.300	445	1027.950
60	138.600	235	542.850	450	1039.500
62	143.220	240	554.400	455	1051.050
64	147.840	245	565.950	460	1062.600
66	152.460	250	577.500	465	1074.150
68	157.080	255	589.050	470	1085.700
70	161.700	260	600.600	475	1097.250
72	166.320	265	612.150	480	1108.800
74	170.940	270	623.700	485	1120.350
76	175.560	275	635.250	490	1131.900

CONVERSION FROM FEET OF HEAD TO PRESSURE (psi)

Head feet	Pressure psi	Head feet	Pressure psi	Head feet	Pressure psi
1	0.433	78	33.774	280	121.240
2	0.866	80	34.640	285	123.405
3	1.299	82	35.506	290	125.570
4	1.732	84	36.372	295	127.735
5	2.165	86	37.238	300	129.900
6	2.598	88	38.104	305	132.065
7	3.031	90	38.970	310	134.230
8	3.464	100	43.300	315	136.395
9	3.897	105	45.465	320	138.560
10	4.330	110	47.630	325	140.725
12	5.196	115	49.795	330	142.890
14	6.062	120	51.960	335	145.055
16	6.928	125	54.125	340	147.220
18	7.794	130	56.290	345	149.385
20	8.660	135	58.455	350	151.550
22	9.526	140	60.620	355	153.715
24	10.392	145	62.785	360	155.880
26	11.258	150	64.950	365	158.045
28	12.124	155	67.115	370	160.210
30	12.990	160	69.280	375	162.375
32	13.856	165	71.445	380	164.540
34	14.722	170	73.610	385	166.705
36	15.588	175	75.775	390	168.870
38	16.454	180	77.940	395	171.035
40	17.320	185	80.105	400	173.200
42	18.186	190	82.270	405	175.365
44	19.052	195	84.435	410	177.530
46	19.918	200	86.600	415	179.695
48	20.784	205	88.765	420	181.860
50	21.650	210	90.930	425	184.025
52	22.516	215	93.095	430	186.190
54	23.382	220	95.260	435	188.355
56	24.248	225	97.425	440	190.520
58	25.114	230	99.590	445	192.685
60	25.980	235	101.755	450	194.850
62	26.846	240	103.920	455	197.015
64	27.712	245	106.085	460	199.180
66	28.578	250	108.250	465	201.345
68	29.444	255	110.415	470	203.510
70	30.310	260	112.580	475	205.675
72	31.176	265	114.745	480	207.840
74	32.042	270	116.910	485	210.005
76	32.908	275	119.075	490	212.170

VOLUMES FOR VERTICAL TANKS (gal./ft.)

Diameter	Volume / Foot Depth	Volume / Foot Depth
feet	cubic feet	gallons
1	0.785	5.876
2	3.142	23.502
3	7.069	52.880
4	12.566	94.009
5	19.635	146.889
6	28.274	211.521
7	38.485	287.903
8	50.266	376.037
9	63.617	475.922
10	78.540	587.558
11	95.033	710.945
12	113.098	846.083
13	132.733	992.973
14	153.938	1151.613
15	176.715	1322.005
16	201.062	1504.148
17	226.981	1698.042
18	254.470	1903.687
19	283.529	2121.083
20	314.160	2350.231
21	346.361	2591.130
22	380.134	2843.779
23	415.477	3108.180
24	452.390	3384.333
25	490.875	3672.236
26	530.930	3971.890
27	572.557	4283.296
28	615.754	4606.453
29	660.521	4941.361
30	706.860	5288.020

APPROXIMATE NUMBER OF SPRINKLERS (per acre)

Square and Rectangular Spacing

Spacing	Heads	Spacing	Heads
feet	*per acre*	*feet*	*per acre*
10 x 10	450	30 x 40	40
11 x 11	375	30 x 50	30
12 x 12	325	30 x 60	25
13 x 13	275	40 x 40	30
14 x 14	225	40 x 50	25
15 x 15	200	40 x 60	20
16 x 16	175	40 x 80	15
17 x 17	150	50 x 50	20
18 x 18	140	50 x 60	15
19 x 19	130	50 x 70	14
20 x 20	120	50 x 80	13
20 x 30	80	60 x 60	12
20 x 40	60	60 x 70	11
20 x 50	50	60 x 80	10
20 x 60	40	70 x 70	9
25 x 25	75	70 x 80	8
30 x 30	55	70 x 90	7

APPROXIMATE NUMBER OF SPRINKLERS (per acre)

Equilateral Triangular Spacing

Spacing feet	Heads per acre	Spacing feet	Heads per acre
10	505	32	50
11	420	34	45
12	350	36	40
13	300	38	35
14	260	40	30
15	225	42	30
16	200	44	25
17	175	46	24
18	155	48	22
19	140	50	20
20	125	52	19
21	115	54	17
22	105	56	16
23	100	58	15
24	90	60	14
25	80	62	13
26	75	64	12
27	70	66	12
28	65	68	11
29	60	70	10
30	55	72	10

PUMP HORSEPOWER REQUIREMENTS (WHP)

Head	Pressure	GPM									
feet	psi	25	50	75	100	150	200	250	300	400	500
10	4.330	0.063	0.126	0.189	0.253	0.379	0.505	0.632	0.758	1.011	1.263
15	6.495	0.095	0.189	0.284	0.379	0.568	0.758	0.947	1.137	1.516	1.895
20	8.660	0.000	0.253	0.379	0.505	0.758	1.011	1.263	1.516	2.021	2.526
25	10.825	0.158	0.316	0.474	0.632	0.947	1.263	1.579	1.895	2.526	3.158
30	12.990	0.189	0.379	0.568	0.758	1.137	1.516	1.895	2.274	3.032	3.789
35	15.155	0.221	0.442	0.663	0.884	1.326	1.768	2.210	2.653	3.537	4.421
40	17.320	0.253	0.505	0.758	1.011	1.516	2.021	2.526	3.032	4.042	5.053
45	19.485	0.284	0.568	0.853	1.137	1.705	2.274	2.842	3.410	4.547	5.684
50	21.650	0.316	0.632	0.947	1.263	1.895	2.526	3.158	3.789	5.053	6.316
60	25.980	0.379	0.758	1.137	1.516	2.274	3.032	3.789	4.547	6.063	7.579
70	30.310	0.442	0.884	1.326	1.768	2.653	3.537	4.421	5.305	7.074	8.842
80	34.640	0.505	1.011	1.516	2.021	3.032	4.042	5.053	6.063	8.084	10.105
90	38.970	0.568	1.137	1.705	2.274	3.410	4.547	5.684	6.821	9.095	11.368
100	43.300	0.632	1.263	1.895	2.526	3.789	5.053	6.316	7.579	10.105	12.631
110	47.630	0.695	1.389	2.084	2.779	4.168	5.558	6.947	8.337	11.116	13.894
120	51.960	0.758	1.516	2.274	3.032	4.547	6.063	7.579	9.095	12.126	15.158
130	56.290	0.821	1.642	2.463	3.284	4.926	6.568	8.210	9.852	13.137	16.421
140	60.620	0.884	1.768	2.653	3.537	5.305	7.074	8.842	10.610	14.147	17.684
150	64.950	0.947	1.895	2.842	3.789	5.684	7.579	9.473	11.368	15.158	18.947
160	69.280	1.011	2.021	3.032	4.042	6.063	8.084	10.105	12.126	16.168	20.210
170	73.610	1.074	2.147	3.221	4.295	6.442	8.589	10.737	12.884	17.179	21.473
180	77.940	1.137	2.274	3.410	4.547	6.821	9.095	11.368	13.642	18.189	22.736
190	82.270	1.200	2.400	3.600	4.800	7.200	9.600	12.000	14.400	19.200	23.999
200	86.600	1.263	2.526	3.789	5.053	7.579	10.105	12.631	15.158	20.210	25.263

PUMP HORSEPOWER REQUIREMENTS (WHP)

Head	Pressure	GPM									
feet	psi	25	50	75	100	150	200	250	300	400	500
210	90.930	1.326	2.653	3.979	5.305	7.958	10.610	13.263	15.915	21.221	26.526
220	95.260	1.389	2.779	4.168	5.558	8.337	11.116	13.894	16.673	22.231	27.789
230	99.590	1.453	2.905	4.358	5.810	8.716	11.621	14.526	17.431	23.242	29.052
240	103.920	1.516	3.032	4.547	6.063	9.095	12.126	15.158	18.189	24.252	30.315
250	108.250	1.579	3.158	4.737	6.316	9.473	12.631	15.789	18.947	25.263	31.578
260	112.580	1.642	3.284	4.926	6.568	9.852	13.137	16.421	19.705	26.273	32.841
270	116.910	1.705	3.410	5.116	6.821	10.231	13.642	17.052	20.463	27.284	34.104
280	121.240	1.768	3.537	5.305	7.074	10.610	14.147	17.684	21.221	28.294	35.368
290	125.570	1.832	3.663	5.495	7.326	10.989	14.652	18.315	21.978	29.305	36.631
300	129.900	1.895	3.789	5.684	7.579	11.368	15.158	18.947	22.736	30.315	37.894
310	134.230	1.958	3.916	5.874	7.831	11.747	15.663	19.578	23.494	31.326	39.157
320	138.560	2.021	4.042	6.063	8.084	12.126	16.168	20.210	24.252	32.336	40.420
330	142.890	2.084	4.168	6.252	8.337	12.505	16.673	20.842	25.010	33.347	41.683
340	147.220	2.147	4.295	6.442	8.589	12.884	17.179	21.473	25.768	34.357	42.946
350	151.550	2.210	4.421	6.631	8.842	13.263	17.684	22.105	26.526	35.368	44.209
360	155.880	2.274	4.547	6.821	9.095	13.642	18.189	22.736	27.284	36.378	45.473
370	160.210	2.337	4.674	7.010	9.347	14.021	18.694	23.368	28.041	37.389	46.736
380	164.540	2.400	4.800	7.200	9.600	14.400	19.200	23.999	28.799	38.399	47.999
390	168.870	2.463	4.926	7.389	9.852	14.779	19.705	24.631	29.557	39.410	49.262
400	173.200	2.526	5.053	7.579	10.105	15.158	20.210	25.263	30.315	40.420	50.525
425	184.025	2.684	5.368	8.052	10.737	16.105	21.473	26.841	32.210	42.946	53.683
450	194.850	2.842	5.684	8.526	11.368	17.052	22.736	28.420	34.104	45.473	56.841
475	205.675	3.000	6.000	9.000	12.000	18.000	23.999	29.999	35.999	47.999	59.999
500	216.500	3.158	6.316	9.473	12.631	18.947	25.263	31.578	37.894	50.525	63.156

VOLTAGE LOSSES FOR ANNEALED COPPER WIRE
(loss per 1000 ft.)

Amperes	14	12	10	8	6	4	2
0.10	0.258	0.162	0.102	0.064	0.040	0.025	0.016
0.20	0.516	0.324	0.204	0.128	0.081	0.051	0.032
0.30	0.774	0.486	0.306	0.192	0.121	0.076	0.048
0.40	1.032	0.648	0.408	0.256	0.161	0.101	0.064
0.50	1.290	0.810	0.510	0.321	0.202	0.127	0.080
0.60	1.548	0.972	0.612	0.385	0.242	0.152	0.095
0.70	1.806	1.134	0.714	0.449	0.282	0.177	0.111
0.80	2.064	1.296	0.816	0.513	0.322	0.202	0.127
0.90	2.322	1.458	0.918	0.577	0.363	0.228	0.143
1.00	2.580	1.620	1.020	0.641	0.403	0.253	0.159
1.10	2.838	1.782	1.122	0.705	0.443	0.278	0.175
1.20	3.096	1.944	1.224	0.769	0.484	0.304	0.191
1.30	3.354	2.106	1.326	0.833	0.524	0.329	0.207
1.40	3.612	2.268	1.428	0.897	0.564	0.354	0.223
1.50	3.870	2.430	1.530	0.962	0.605	0.380	0.239
1.60	4.128	2.592	1.632	1.026	0.645	0.405	0.254
1.70	4.386	2.754	1.734	1.090	0.685	0.430	0.270
1.80	4.644	2.916	1.836	1.154	0.725	0.455	0.286
1.90	4.902	3.078	1.938	1.218	0.766	0.481	0.302
2.00	5.160	3.240	2.040	1.282	0.806	0.506	0.318
2.50	6.450	4.050	2.550	1.603	1.008	0.633	0.398
3.00	7.740	4.860	3.060	1.923	1.209	0.759	0.477
3.50	9.030	5.670	3.570	2.244	1.411	0.886	0.557
4.00	10.320	6.480	4.080	2.564	1.612	1.012	0.636
4.50	11.610	7.290	4.590	2.885	1.814	1.139	0.716
5.00	12.900	8.100	5.100	3.205	2.015	1.265	0.795
5.50	14.190	8.910	5.610	3.526	2.217	1.392	0.875
6.00	15.480	9.720	6.120	3.846	2.418	1.518	0.954
6.50	16.770	10.530	6.630	4.167	2.620	1.645	1.034
7.00	18.060	11.340	7.140	4.487	2.821	1.771	1.113
7.50	19.350	12.150	7.650	4.808	3.023	1.898	1.193

* Losses calculated at 25 degrees Celsius and 77 degrees Fahrenheit.

** Voltage losses are calculated using the following formula:

$V = IR$

where I = current in amperes

where R = resistance in ohms/1000 ft.

where V = voltage

*** Resistance in ohms/1000 ft. are as follows:

14 gauge=	2.58	6 gauge=	0.403
12 gauge=	1.62	4 gauge=	0.253
10 gauge=	1.02	2 gauge=	0.159
8 gauge=	0.641		

DECIMAL EQUIVALENTS OF INCHES AND FEET

	Decimals of an Inch				Decimals of a Foot		
Fractions of an inch	Decimals of an inch	Fractions of an inch	Decimals of an inch	Fraction in inches	Decimals of a foot	Fraction in inches	Decimals of a foot
0	0.000000	1/2	0.500000	0	0.000000	4 1/4	0.354167
1/64	0.015625	33/64	0.515625	1/64	0.001302	4 1/2	0.375000
1/32	0.031250	17/32	0.531250	1/32	0.002604	4 3/4	0.395833
3/64	0.046875	35/64	0.546875	1/16	0.005208	5	0.416667
1/16	0.062500	9/16	0.562500	3/32	0.007813	5 1/4	0.437500
5/64	0.078125	37/64	0.578125	1/8	0.010417	5 1/2	0.458333
3/32	0.093750	19/32	0.593750	3/16	0.015625	5 3/4	0.479167
7/64	0.109375	39/64	0.609375	1/4	0.020833	6	0.500000
1/8	0.125000	5/8	0.625000	5/16	0.026042	6 1/4	0.520833
9/64	0.140625	41/64	0.640625	3/8	0.031250	6 1/2	0.541667
5/32	0.156250	21/32	0.656250	7/16	0.036458	6 3/4	0.562500
11/64	0.171875	43/64	0.671875	1/2	0.041667	7	0.583333
3/16	0.187500	11/16	0.687500	9/16	0.046875	7 1/4	0.604167
13/64	0.203125	45/64	0.703125	5/8	0.052083	7 1/2	0.625000
7/32	0.218750	23/32	0.718750	11/16	0.057292	7 3/4	0.645833
15/64	0.234375	47/64	0.734375	3/4	0.062500	8	0.666667
1/4	0.250000	3/4	0.750000	13/16	0.067708	8 1/4	0.687500
17/64	0.265625	49/64	0.765625	7/8	0.072917	8 1/2	0.708333
9/32	0.281250	25/32	0.781250	15/16	0.078125	8 3/4	0.729167
19/64	0.296875	51/64	0.796875	1	0.083333	9	0.750000
5/16	0.312500	13/16	0.812500	1 1/4	0.104167	9 1/4	0.770833
21/64	0.328125	53/64	0.828125	1 1/2	0.125000	9 1/2	0.791667
11/32	0.343750	27/32	0.843750	1 3/4	0.145833	9 3/4	0.812500
23/64	0.359375	55/64	0.859375	2	0.166667	10	0.833333
3/8	0.375000	7/8	0.875000	2 1/4	0.187500	10 1/4	0.854167
25/64	0.390625	57/64	0.890625	2 1/2	0.208333	10 1/2	0.875000
13/32	0.406250	29/32	0.906250	2 3/4	0.229167	10 3/4	0.895833
27/64	0.421875	59/64	0.921875	3	0.250000	11	0.916667
7/16	0.437500	15/16	0.937500	3 1/4	0.270833	11 1/4	0.937500
29/64	0.453125	61/64	0.953125	3 1/2	0.291667	11 1/2	0.958333
15/32	0.468750	31/32	0.968750	3 3/4	0.312500	11 3/4	0.979167
31/64	0.484375	63/64	0.984375	4	0.333333	12	1.000000

METRIC CONVERSIONS

Inches	Millimeters	Feet	Meters	Feet	Meters	Feet	Meters
1/128	0.1984	1	0.305	35	10.668	69	21.031
1/64	0.3969	2	0.610	36	10.973	70	21.336
3/128	0.5953	3	0.914	37	11.278	71	21.641
1/32	0.7938	4	1.219	38	11.582	72	21.946
5/128	0.9922	5	1.524	39	11.887	73	22.250
3/64	1.1906	6	1.829	40	12.192	74	22.555
7/128	1.3891	7	2.134	41	12.497	75	22.860
1/16	1.5875	8	2.438	42	12.802	76	23.165
1/8	3.1750	9	2.743	43	13.106	77	23.470
3/16	4.7625	10	3.048	44	13.411	78	23.774
1/4	6.3500	11	3.353	45	13.716	79	24.079
5/16	7.9375	12	3.658	46	14.021	80	24.384
3/8	9.5250	13	3.962	47	14.326	81	24.689
7/16	11.1125	14	4.267	48	14.630	82	24.994
1/2	12.7000	15	4.572	49	14.935	83	25.298
9/16	14.2875	16	4.877	50	15.240	84	25.603
5/8	15.8750	17	5.182	51	15.545	85	25.908
11/16	17.4625	18	5.486	52	15.850	86	26.213
3/4	19.0500	19	5.791	53	16.154	87	26.518
13/16	20.6375	20	6.096	54	16.459	88	26.822
7/8	22.2250	21	6.401	55	16.764	89	27.127
15/16	23.8125	22	6.706	56	17.069	90	27.432
1	25.4001	23	7.010	57	17.374	91	27.737
2	50.8001	24	7.315	58	17.678	92	28.042
3	76.2002	25	7.620	59	17.983	93	28.346
4	101.6002	26	7.925	60	18.288	94	28.651
5	127.0003	27	8.230	61	18.593	95	28.956
6	152.4003	28	8.534	62	18.898	96	29.261
7	177.8004	29	8.839	63	19.202	97	29.566
8	203.2004	30	9.144	64	19.507	98	29.870
9	228.6005	31	9.449	65	19.812	99	30.175
10	254.0005	32	9.754	66	20.117	100	30.480
11	279.4006	33	10.058	67	20.422		
12	304.8006	34	10.363	68	20.726		

AREA OF CIRCLES

Diameter	Area	Diameter	Area	Diameter	Area
feet	sq. feet	feet	sq. feet	feet	sq. feet
1	0.79	78	4778	260	53,093
2	3.14	80	5027	265	55,155
3	7.07	82	5281	270	57,256
4	12.57	84	5542	275	59,396
5	19.64	86	5809	280	61,575
6	28.27	88	6082	285	63,794
7	38.48	90	6362	290	66,052
8	50.27	92	6648	295	68,349
9	63.62	94	6940	300	70,686
10	78.54	96	7238	305	73,062
12	113.10	98	7543	310	75,477
14	153.94	100	7854	315	77,931
16	201.06	105	8659	320	80,425
18	254.47	110	9503	325	82,958
20	314.16	115	10,387	330	85,530
22	380.13	120	11,310	335	88,142
24	452.39	125	12,272	340	90,792
26	530.93	130	13,273	345	93,482
28	615.75	135	14,314	350	96,212
30	706.86	140	15,394	355	98,980
32	804.25	145	16,513	360	101,788
34	907.92	150	17,672	365	104,635
36	1018	155	18,869	370	107,521
38	1134	160	20,106	375	110,447
40	1257	165	21,383	380	113,412
42	1385	170	22,698	385	116,416
44	1521	175	24,053	390	119,459
46	1662	180	25,447	395	122,542
48	1810	185	26,880	400	125,664
50	1964	190	28,353	405	128,825
52	2124	195	29,865	410	132,026
54	2290	200	31,416	415	135,266
56	2463	205	33,006	420	138,545
58	2642	210	34,636	425	141,863
60	2827	215	36,305	430	145,220
62	3019	220	38,013	435	148,617
64	3217	225	39,761	440	152,053
66	3421	230	41,548	445	155,529
68	3632	235	43,374	450	159,044
70	3848	240	45,239	455	162,597
72	4072	245	47,144	460	166,191
74	4301	250	49,088	465	169,823
76	4536	255	51,071	470	173,495

RESISTANCE IN ANNEALED COPPER WIRES

Wire Gauge	Wire Type	Resistance in Ohms per 1000 Ft.	
(AWG)	(UF)	77 F	149 F
18	Solid	6.51	7.51
16	Solid	4.09	4.73
14	Solid	2.58	2.97
12	Solid	1.62	1.87
10	Solid	1.02	1.18
8	Solid	0.641	0.739
6	Stranded	0.403	0.465
4	Stranded	0.253	0.292
2	Stranded	0.159	0.184
0	Stranded	0.100	0.116

NUMBER OF WIRES INSTALLED IN PVC SLEEVING

Wire Size (AWG)	Approximate Number of Wires in Sleeving											
	1/2"	3/4"	1"	1 1/4"	1 1/2"	2"	2 1/2"	3"	3 1/2"	4"	5"	6"
16	5	10	15	30	40	65	95	140	-	-	-	-
14	4	6	10	17	22	35	55	80	110	140	-	-
12	3	5	6	15	18	30	45	70	100	120	200	-
10	1	4	5	12	15	25	35	60	80	100	160	-
8	1	2	4	5	7	15	25	30	45	60	100	140
6	1	1	3	3	4	8	12	20	30	35	60	85
4		1	1	2	3	5	8	15	20	25	45	60
2		1	1	2	2	4	7	10	15	20	30	45

WATER USE ANALYSIS WORKSHEET

Mill Creek Home Owner's Association

11/30/94

*Figures are in thousand gallons

METER LOCATION	6848 S. Revere Parkway			6868 S. Revere Parkway		
	1992	1993	1994	1992	1993	1994
JAN	0	0	0	0	0	0
FEB	0	0	0	0	0	0
MAR	0	0	0	0	0	0
APR	6	30	0	31	36	0
MAY	79	136	74	99	162	70
JUN	92	152	181	132	382	170
JUL	167	110	193	188	360	368
AUG	257	94	281	321	189	435
SEP	254	147	209	293	169	374
OCT	47	109	143	57	74	215
NOV	0	0	0	0	0	0
DEC	0	0	0	7	0	0
TOTAL USE	902	778	1,081	1,121	1,372	1,632

	1992	1993	1994
IRRIGATION USE:	2,023	2,150	2,713
AREA IN SQ. FT.	63,608		
ANNUAL INCHES/SQ.FT.	51.02	54.22	68.42
ANNUAL WATER COST AT THE 1993 RATE OF **$3.50** per THOUSAND GALLONS:	**$7,081**	**$7,525**	**$9,496**
POTENTIAL WATER SAVINGS BASED ON:	**26**	INCHES ANNUALLY	
	$3,472	**$3,917**	**$5,887**

CAD DESIGN DRAWINGS

ELECTRIC CONTROL VALVE DETAIL

ALL STAINLESS STEEL LOCKING
PINCH CLAMPS (USE 2 CLAMPS FOR
PIPE 1 1/2" AND LARGER)

BRAND VALVE BOX LID WITH CONTROL
VALVE # AS SHOWN ON PLAN.

12" RECTANGULAR VALVE
BOX, REFER TO DETAIL

SOLENOID VALVE

RESILIENT SEAT GATE
VALVE, LINE SIZE

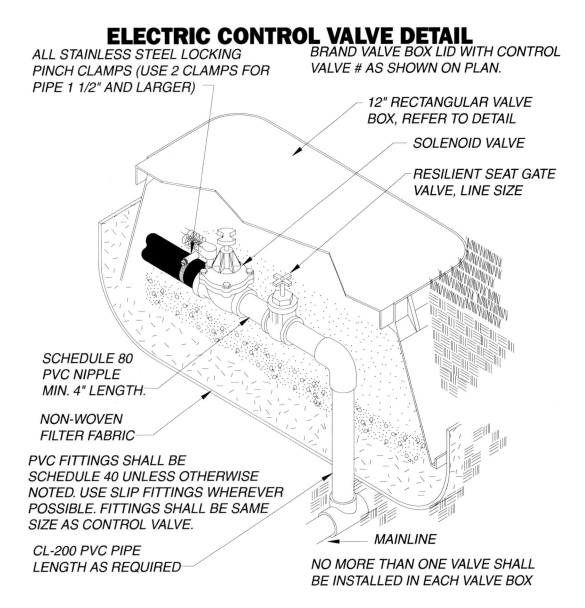

SCHEDULE 80
PVC NIPPLE
MIN. 4" LENGTH.

NON-WOVEN
FILTER FABRIC

PVC FITTINGS SHALL BE
SCHEDULE 40 UNLESS OTHERWISE
NOTED. USE SLIP FITTINGS WHEREVER
POSSIBLE. FITTINGS SHALL BE SAME
SIZE AS CONTROL VALVE.

CL-200 PVC PIPE
LENGTH AS REQUIRED

MAINLINE

NO MORE THAN ONE VALVE SHALL
BE INSTALLED IN EACH VALVE BOX

©Copyright Keesen Water Management, Inc. 1993
Denver, Colorado

NOT TO SCALE

BACKFLOW PREVENTER DETAIL

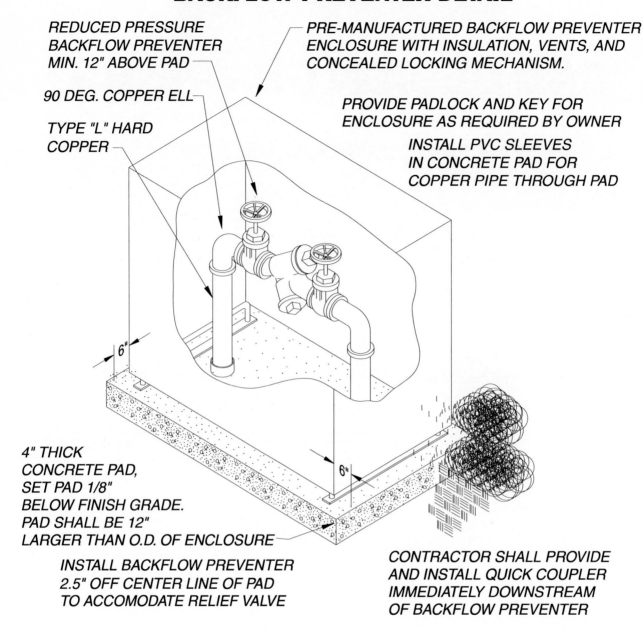

REDUCED PRESSURE
BACKFLOW PREVENTER
MIN. 12" ABOVE PAD

PRE-MANUFACTURED BACKFLOW PREVENTER
ENCLOSURE WITH INSULATION, VENTS, AND
CONCEALED LOCKING MECHANISM.

90 DEG. COPPER ELL

PROVIDE PADLOCK AND KEY FOR
ENCLOSURE AS REQUIRED BY OWNER

TYPE "L" HARD
COPPER

INSTALL PVC SLEEVES
IN CONCRETE PAD FOR
COPPER PIPE THROUGH PAD

6"

4" THICK
CONCRETE PAD,
SET PAD 1/8"
BELOW FINISH GRADE.
PAD SHALL BE 12"
LARGER THAN O.D. OF ENCLOSURE

6"

INSTALL BACKFLOW PREVENTER
2.5" OFF CENTER LINE OF PAD
TO ACCOMODATE RELIEF VALVE

CONTRACTOR SHALL PROVIDE
AND INSTALL QUICK COUPLER
IMMEDIATELY DOWNSTREAM
OF BACKFLOW PREVENTER

©Copyright Keesen Water Management, Inc. 1993
Denver, Colorado

NOT TO SCALE

CONTROLLER DETAIL

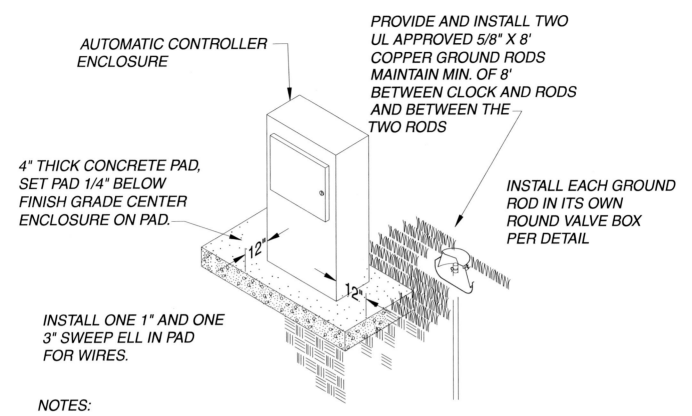

AUTOMATIC CONTROLLER ENCLOSURE

PROVIDE AND INSTALL TWO UL APPROVED 5/8" X 8' COPPER GROUND RODS MAINTAIN MIN. OF 8' BETWEEN CLOCK AND RODS AND BETWEEN THE TWO RODS

4" THICK CONCRETE PAD, SET PAD 1/4" BELOW FINISH GRADE CENTER ENCLOSURE ON PAD.

INSTALL EACH GROUND ROD IN ITS OWN ROUND VALVE BOX PER DETAIL

12"

12"

INSTALL ONE 1" AND ONE 3" SWEEP ELL IN PAD FOR WIRES.

NOTES:

CONTROLLER SIZE, LOCATION AND TYPE AS SHOWN ON IRRIGATION PLANS.

CONTROLLER PEDESTAL TO BE APPROPRIATE PEDESTAL AS RECOMMENDED BY CONTROLLER MANUFACTURER, UNLESS OTHERWISE INDICATED ON PLANS.

GROUND WIRE TO BE #8 COPPER WIRE CONNECTING WITHIN THE TOP 4" OF GROUND ROD. CONNECT TO ROD WITH UL APPROVED GROUND CLAMP (NO AUTOMOBILE OR SCREW CLAMPS)

EACH CONTROLLER IS TO HAVE TWO SEPARATE GROUND RODS, ONE FOR PRIMARY SURGE PROTECTION AND ONE FOR SECONDARY SURGE PROTECTION

CONTROLLERS SHALL BE INSTALLED WITH MANUFACTURER'S SECONDARY SURGE PROTECTION AND PRIMARY LIGHTNING ARRESTOR

NOT TO SCALE

EMITTER INSTALLATION DETAIL

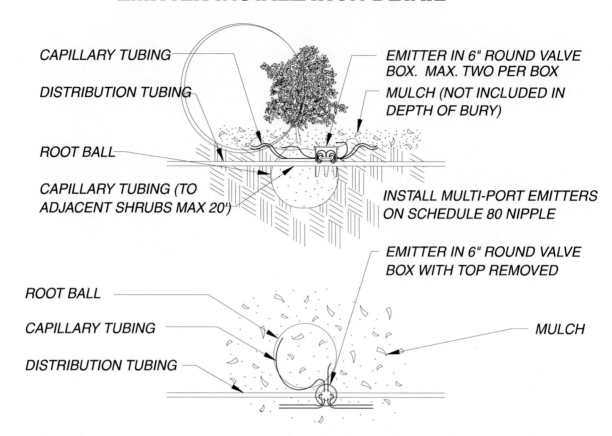

CAPILLARY TUBING

DISTRIBUTION TUBING

ROOT BALL

CAPILLARY TUBING (TO ADJACENT SHRUBS MAX 20')

EMITTER IN 6" ROUND VALVE BOX. MAX. TWO PER BOX

MULCH (NOT INCLUDED IN DEPTH OF BURY)

INSTALL MULTI-PORT EMITTERS ON SCHEDULE 80 NIPPLE

EMITTER IN 6" ROUND VALVE BOX WITH TOP REMOVED

ROOT BALL

CAPILLARY TUBING

DISTRIBUTION TUBING

MULCH

NOTES:
-MULCH DEPTH SHALL NOT BE INCLUDED IN DEPTHS OF BURY, ALL TUBING MUST BE BURIED BY THE INDICATED AMOUNT OF SOIL IN ADDITION TO MULCH. REFER TO PLAN SHEETS FOR REQUIRED DEPTH OF BURY FOR DRIP DISTRIBUTION PIPE. ALL CAPILLARY TUBING MUST BE BURIED BENEATH 2" OF SOIL, OUTLET SHALL BE STAKED WITH DRIP MANUFACTURER'S STAKE 1/2" ABOVE FINISH GRADE.
-EMISSION POINTS ARE TO BE EQUALLY SPACED AROUND PLANT, FOR SLOPE APPLICATIONS EMITTERS SHALL BE PLACED ON UPHILL SIDE OF PLANT, EMITTERS SHALL BE ABOVE ROOT BALL 2" INSIDE ROOT BALL EDGE
-USE ONE 1/2 GPH EMISSION POINT FOR PERENNIALS, GROUNDCOVER, & SHRUBS SMALLER THAN 5 GALLON, USE TWO 1/2 GPH EMISSION POINTS FOR 5 GALLON AND LARGER SHRUBS, USE SIX 1 GPH FOR EACH TREE.
-DO NOT DRIP IRRIGATE ANY PLANT MATERIAL IN AREAS WHICH ARE SPRAY IRRIGATED

NOT TO SCALE

ELECTRIC CONTROL VALVE DETAIL

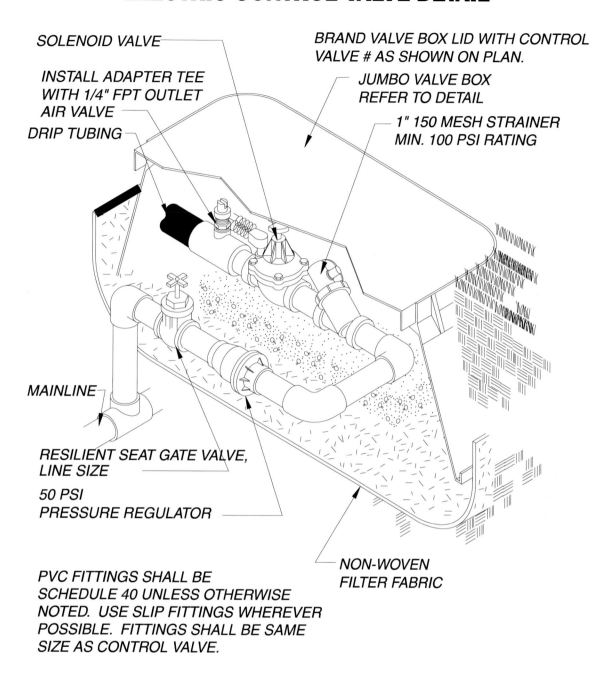

SOLENOID VALVE

INSTALL ADAPTER TEE
WITH 1/4" FPT OUTLET
AIR VALVE

DRIP TUBING

BRAND VALVE BOX LID WITH CONTROL
VALVE # AS SHOWN ON PLAN.

JUMBO VALVE BOX
REFER TO DETAIL

1" 150 MESH STRAINER
MIN. 100 PSI RATING

MAINLINE

RESILIENT SEAT GATE VALVE,
LINE SIZE

50 PSI
PRESSURE REGULATOR

NON-WOVEN
FILTER FABRIC

PVC FITTINGS SHALL BE
SCHEDULE 40 UNLESS OTHERWISE
NOTED. USE SLIP FITTINGS WHEREVER
POSSIBLE. FITTINGS SHALL BE SAME
SIZE AS CONTROL VALVE.

NOT TO SCALE

FLUSH VALVE DETAIL

BRAND VALVE BOX LID WITH "F.V."

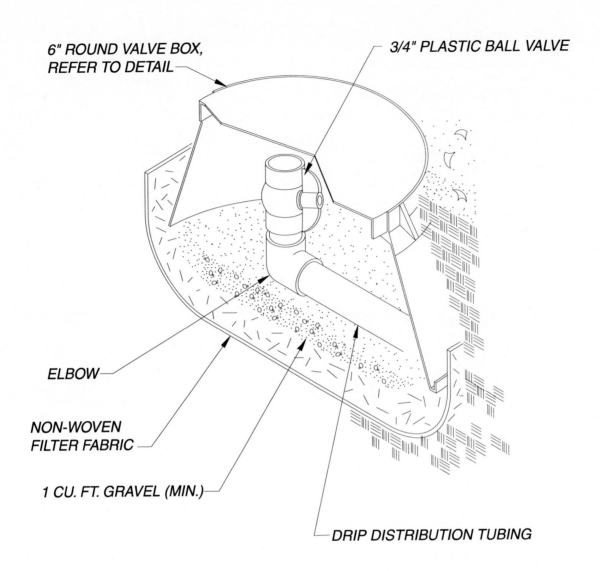

6" ROUND VALVE BOX,
REFER TO DETAIL

3/4" PLASTIC BALL VALVE

ELBOW

NON-WOVEN
FILTER FABRIC

1 CU. FT. GRAVEL (MIN.)

DRIP DISTRIBUTION TUBING

©Copyright Keesen Water Management, Inc. 1993
Denver, Colorado

NOT TO SCALE

ISOLATION VALVE DETAIL

10" ROUND VALVE BOX
REFER TO DETAIL

BRAND VALVE BOX LID WITH "I.V."

NON-WOVEN
FILTER FABRIC

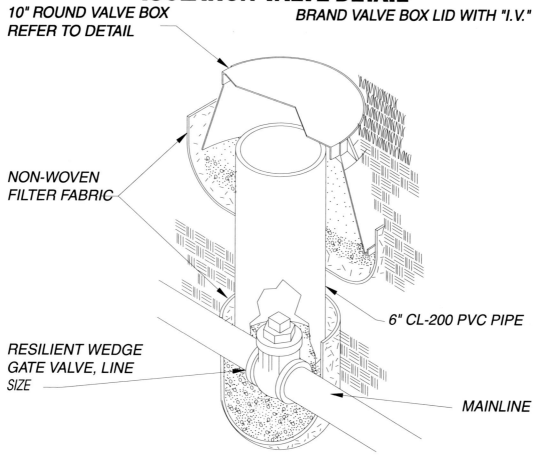

6" CL-200 PVC PIPE

RESILIENT WEDGE
GATE VALVE, LINE
SIZE

MAINLINE

NOT TO SCALE

MANUAL DRAIN VALVE DETAIL

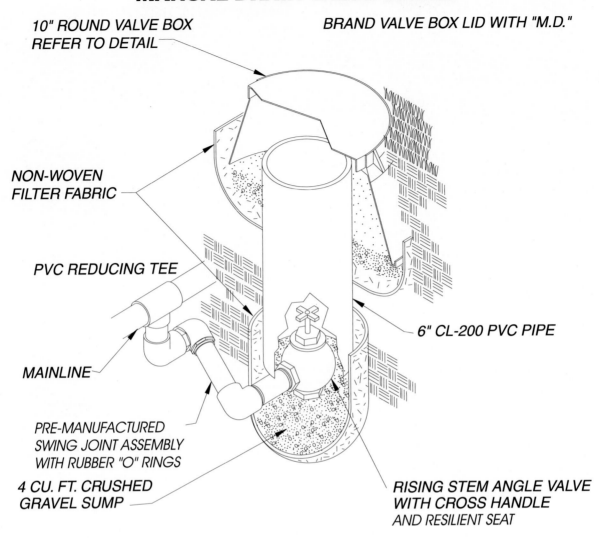

10" ROUND VALVE BOX
REFER TO DETAIL

BRAND VALVE BOX LID WITH "M.D."

NON-WOVEN
FILTER FABRIC

PVC REDUCING TEE

MAINLINE

6" CL-200 PVC PIPE

PRE-MANUFACTURED
SWING JOINT ASSEMBLY
WITH RUBBER "O" RINGS

4 CU. FT. CRUSHED
GRAVEL SUMP

RISING STEM ANGLE VALVE
WITH CROSS HANDLE
AND RESILIENT SEAT

NOTE: DRAIN VALVE MUST BE
INSTALLED BELOW MAINLINE

NOT TO SCALE

POINT OF CONNECTION DETAIL

BRAND VALVE BOX LID WITH "P.O.C."

10" ROUND VALVE BOX
REFER TO DETAIL

NON-WOVEN
FILTER FABRIC

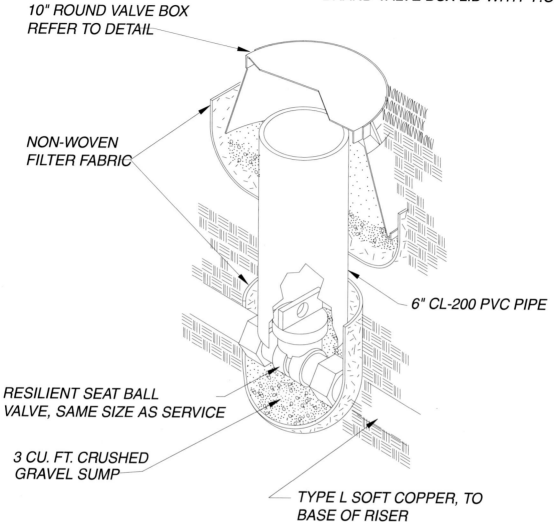

6" CL-200 PVC PIPE

RESILIENT SEAT BALL
VALVE, SAME SIZE AS SERVICE

3 CU. FT. CRUSHED
GRAVEL SUMP

TYPE L SOFT COPPER, TO
BASE OF RISER

©Copyright Keesen Water Management, Inc. 1993
Denver, Colorado

NOT TO SCALE

PRESSURE REDUCING VALVE DETAIL

BRAND VALVE BOX LID "P.R.V."

PRESSURE REDUCING VALVE

JUMBO VALVE BOX
REFER TO DETAIL

INSTALL ADAPTER TEE
WITH 1/4" FPT OUTLET
AIR VALVE

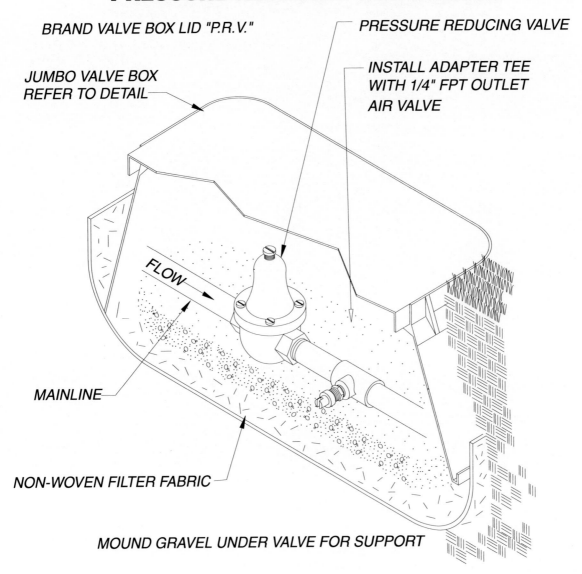

FLOW

MAINLINE

NON-WOVEN FILTER FABRIC

MOUND GRAVEL UNDER VALVE FOR SUPPORT

NOT TO SCALE

POP-UP SPRAY HEAD DETAIL

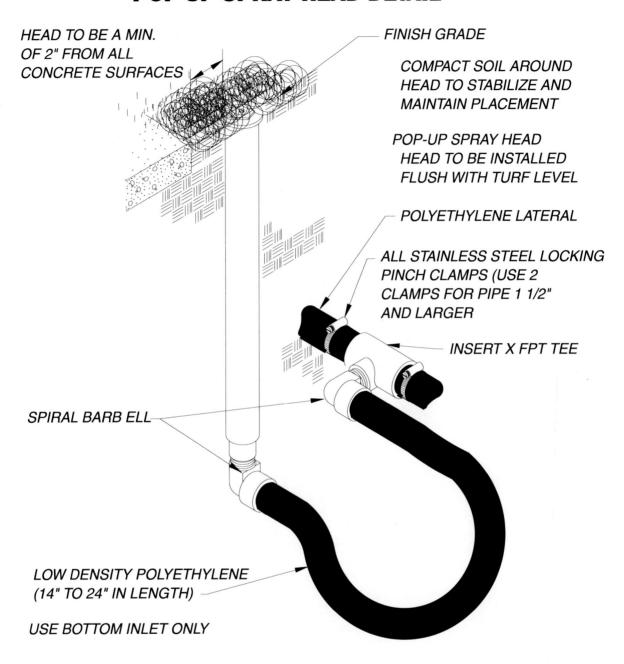

HEAD TO BE A MIN.
OF 2" FROM ALL
CONCRETE SURFACES

FINISH GRADE

COMPACT SOIL AROUND
HEAD TO STABILIZE AND
MAINTAIN PLACEMENT

POP-UP SPRAY HEAD
HEAD TO BE INSTALLED
FLUSH WITH TURF LEVEL

POLYETHYLENE LATERAL

ALL STAINLESS STEEL LOCKING
PINCH CLAMPS (USE 2
CLAMPS FOR PIPE 1 1/2"
AND LARGER

INSERT X FPT TEE

SPIRAL BARB ELL

LOW DENSITY POLYETHYLENE
(14" TO 24" IN LENGTH)

USE BOTTOM INLET ONLY

NOT TO SCALE

POP-UP SPRAY HEAD DETAIL

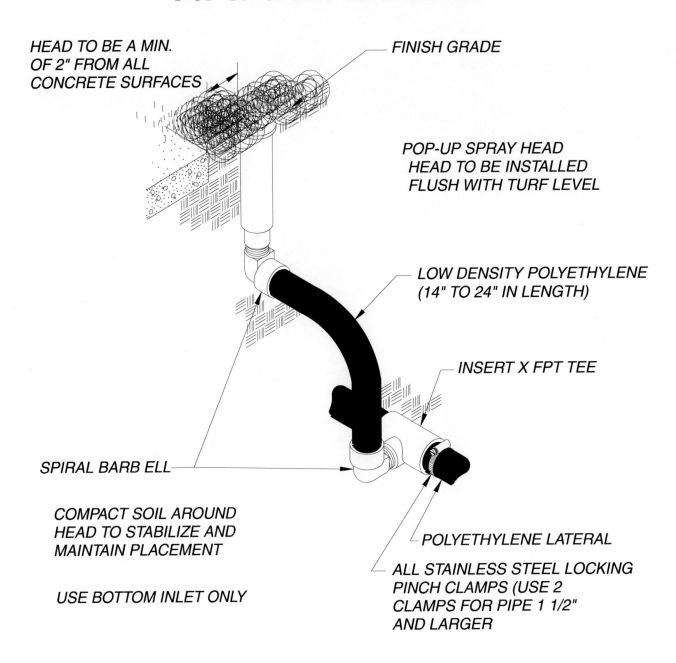

HEAD TO BE A MIN.
OF 2" FROM ALL
CONCRETE SURFACES

FINISH GRADE

POP-UP SPRAY HEAD
HEAD TO BE INSTALLED
FLUSH WITH TURF LEVEL

LOW DENSITY POLYETHYLENE
(14" TO 24" IN LENGTH)

INSERT X FPT TEE

SPIRAL BARB ELL

COMPACT SOIL AROUND
HEAD TO STABILIZE AND
MAINTAIN PLACEMENT

USE BOTTOM INLET ONLY

POLYETHYLENE LATERAL

ALL STAINLESS STEEL LOCKING
PINCH CLAMPS (USE 2
CLAMPS FOR PIPE 1 1/2"
AND LARGER

NOT TO SCALE

QUICK COUPLER VAVLE DETAIL

BRAND VALVE BOX LID WITH "Q.C."

10" ROUND VALVE
BOX, REFER TO DETAIL

NON-WOVEN
FILTER FABRIC

TWO PIECE QUICK
COUPLING VALVE

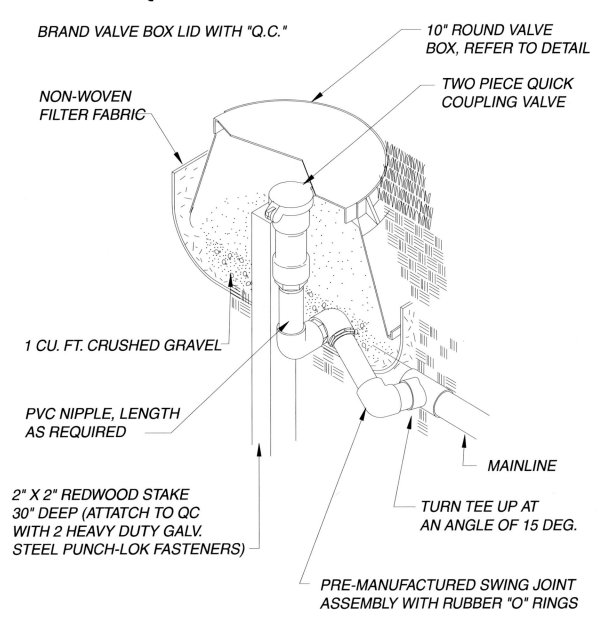

1 CU. FT. CRUSHED GRAVEL

PVC NIPPLE, LENGTH
AS REQUIRED

MAINLINE

2" X 2" REDWOOD STAKE
30" DEEP (ATTATCH TO QC
WITH 2 HEAVY DUTY GALV.
STEEL PUNCH-LOK FASTENERS)

TURN TEE UP AT
AN ANGLE OF 15 DEG.

PRE-MANUFACTURED SWING JOINT
ASSEMBLY WITH RUBBER "O" RINGS

©Copyright Keesen Water Management, Inc. 1993
Denver, Colorado

NOT TO SCALE

ROTARY SPRAY HEAD DETAIL

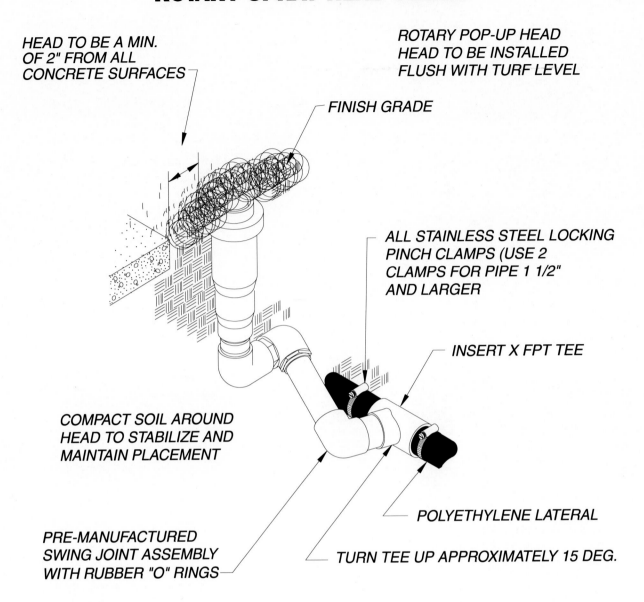

HEAD TO BE A MIN.
OF 2" FROM ALL
CONCRETE SURFACES

ROTARY POP-UP HEAD
HEAD TO BE INSTALLED
FLUSH WITH TURF LEVEL

FINISH GRADE

ALL STAINLESS STEEL LOCKING
PINCH CLAMPS (USE 2
CLAMPS FOR PIPE 1 1/2"
AND LARGER

INSERT X FPT TEE

COMPACT SOIL AROUND
HEAD TO STABILIZE AND
MAINTAIN PLACEMENT

POLYETHYLENE LATERAL

PRE-MANUFACTURED
SWING JOINT ASSEMBLY
WITH RUBBER "O" RINGS

TURN TEE UP APPROXIMATELY 15 DEG.

NOT TO SCALE

SLOPE CONVERSION GRAPHIC

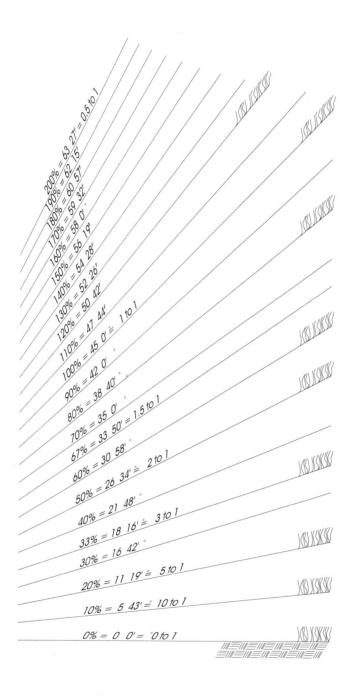

200% = 63 27' = 0.5 to 1
190% = 62 15'
180% = 59 32'
170% = 58 0'
160% = 56 19'
150% = 54 28'
140% = 52 26'
130% = 50 42'
120% = 47 44' = 1 to 1
110% = 45 0'
100% = 42 0'
90% = 38 40'
80% = 35 0'
70% = 33 50' = 1.5 to 1
67% = 30 58'
60% = 26 34' = 2 to 1
50% = 21 48'
40% = 18 16' = 3 to 1
33% = 16 42'
30% = 11 19' = 5 to 1
20% = 5 43' = 10 to 1
10% = 0 0' = 0 to 1
0% = 0 0' = 0 to 1

GRAPHIC SHOWS EQUIVALENT MEASURE OF SLOPE IN:
PER CENT, DEGREES MINUTES, & RATIO

NOT TO SCALE

MAINLINE & WIRING DETAIL

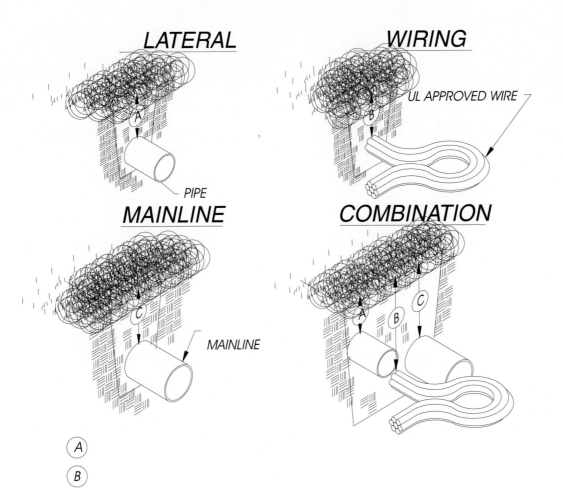

LATERAL

A

PIPE

WIRING

B

UL APPROVED WIRE

MAINLINE

C

MAINLINE

COMBINATION

A   B   C

(A)
(B)
(C)

SNAKE ALL PIPE IN TRENCHES PRIOR TO BACKFILL.
PROVIDE A 24" EXPANSION LOOP FOR WIRE WHENEVER
CHANGE OF DIRECTION IS GREATER THAN 40 DEG. OR LENGTH
EXCEEDS 300' (INSTALL WIRE
CONNECTIONS IN 10" ROUND VALVE BOXES). TAPE AND
BUNDLE WIRE AT 15' INTERVALS. USE 3MDBY, OR 3M DBR
SPLICE KIT OR APPROVED EQUAL FOR WIRE CONNECTIONS.

NOT TO SCALE

VALVE BOX DETAIL

INSTALL ALL VALVE BOXES
APPROX. 1/8" BELOW
FINISH GRADE.

BRAND ALL VALVE BOX LIDS WITH
IDENTIFIICATION AS INDICATED ON
INDIVIDUAL DETAILS.

HDPE VALVE BOX
AND EXTENSION(S) OR
APPROVED EQUAL.

MAINTAIN A MINIMUM OF 1"
BETWEEN VALVE BOX LID AND
TOP OF FLOW CONTROL
HANDLE IN OPEN POSITION

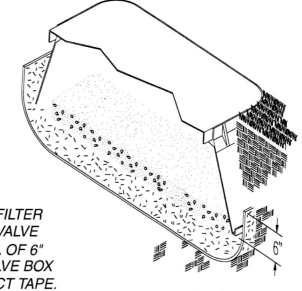

INSTALL NON-WOVEN FILTER
FABRIC BENEATH ALL VALVE
BOXES, EXTEND A MIN. OF 6"
FROM BOTTOM OF VALVE BOX
AND ATTACH WITH DUCT TAPE.

INSTALL A MIN. OF ONE 12" EXTENSION
FOR EACH VALVE BOX, ADDITIONAL
EXTENSIONS AS REQUIRED.

INSTALL 3/4" CRUSHED GRAVEL
TO WITHIN 1" OF BOTTOM OF VALVE.

VALVE BOXES SHALL BE ALIGNED
WITH ADJACENT SURFACES, OTHER
VALVE BOXES AND GRADE CONTOURS.

BACKFILL AROUND VALVE BOX IN
6" LIFTS AND COMPACT SOIL, REPEAT
BACKFILL IN 6" INCREMENTS
UNTIL BACKFILL IS COMPLETE.

DO NOT INSTALL VALVE BOXES
AT LOW POINTS.

NOT TO SCALE

GLOSSARY

Anemometer A spinning device with cups mounted on a pole that measures wind speed.

Application Rate The rate at which water is applied to the landscape by the irrigation system, and measured in inches applied per hour.

Arc of Coverage The portion or degrees of a circle wetted by a single sprinkler head. A 90-degree arc represents a quarter circle, a 120-degree arc is a third of a circle and a 180-degree arc is a half circle.

As-Built Drawing A record of the location, with dimensions, of the installed irrigation system equipment. Changes in sprinkler head and lateral line locations are shown, but not dimensioned. As-built drawings are provided by the installer.

Atmospheric Pressure The pressure exerted in every direction at any given point by the weight of the Earth's atmosphere. Increased altitude and storms will cause a change in atmospheric pressure. At sea level atmospheric pressure is equivalent to 14.7 PSI or 33.9 feet of head and the barometric pressure is 29.9 inches of mercury. At an altitude of 6,000 feet the atmospheric pressure is 11.8 PSI or 27.2 feet of head and the barometric pressure is 24 inches. Atmospheric pressure can change when a storm occurs, while pressure can change plus or minus 15 percent of normal conditions.

Available Water Holding Capacity (AWHC) The water holding capacity or field capacity of a soil type is defined as the amount of water that remains in the soil after the gravitational water has drained away and after the rate of downward movement has decreased to a significant extent. It varies with soil type.

Backflow Preventer A mechanical device installed at the beginning of the irrigation system to prevent a reverse water flow and possible contamination of the potable water supply. Reduced pressure backflow preventers, pressure vacuum breakers and double-check valves are the most common types used in the irrigation industry.

Cavitation The rapid creation and disintegration of vapor bubbles that occurs when the pressure on the water is less than the vapor pressure. Damage may occur in a centrifugal pump at the leading edge of the impeller vane tips. When water enters the eye of the impeller an increase in velocity takes place accompanied by lower water pressure.

Check Valve A device that prevents a reverse water flow in pipes and sprinkler heads. Commonly used in sprinkler heads to prevent low head drainage.

Control Valve An electrically actuated valve used to operate each zone of the irrigation system.

Controller An electric timing device that operates each zone for a predetermined time and frequency.

Coverage The extent of irrigation coverage provided by sprinkler heads with respect to proper head spacing.

Crop Coefficient ET is usually based on a field crop such as alfalfa. A crop coefficient or percentage is used to determine the specific plant water requirement.

Cycle An interval of time during which the irrigation system is operated for one sequence of a regularly recurring succession of watering events. Also referred to as "repeat cycles" and used to prevent runoff.

Design Used as a basis for anticipating practical problems and solving them at the planning stage. Also used to create a drawing showing the location, type and size of the irrigation system components prior to installation.

Distribution Uniformity A method of calculating system uniformity — measured as a percentage — which indicates how closely the driest area compares to the average precipitation rate of the irrigated area.

Drain Valve A valve used to drain the irrigation system for repairs or freeze protection. Referred to as manual drains or automatic drains.

Drip Emitter A small device emitting a dribble of water at a very low pressure with flows measured in gallons per hour.

Dynamic Pressure The pressure at any point in the irrigation system when the system is functioning. Measured with flow in the line.

Evaporation The change by which water is converted from a liquid state into vapor and carried off into the atmosphere.

Evapotranspiration (ET) The amount of water lost from the soil both by surface evaporation and water used by the plant for cooling the leaf surface (transpiration). This process enables the plant to maintain its structure.

Field Capacity Maximum water holding capacity of the soil which is usually measured in inches of water per inch of soil or soil depth.

Flow Sensor A device that measures the rate of water flow in the system in gallons per minute (GPM).

Friction Loss Water rubbing against the system components resulting in accumulated pressure loss from the source to the nozzle.

Friction Head Pipe friction or pressure loss measured in head of feet instead of PSI.

Gallons Per Minute (GPM) A measurement of the quantity and flow rate of water.

Head Pumping design is always calculated in feet of head instead of PSI (one foot of head equals 0.433 PSI), because of the ease of use and precise calculations required.

Infiltration Rate The rate in inches per hour that water moves into the soil. This rate slows with time.

Irrigation Efficiency The percentage of water applied that is actually used by the plant.

Lateral Piping between the control valve and the sprinkler head.

Low Head Drainage A condition in which water drains partially or completely out of the lateral line through the sprinkler head after each irrigation cycle is completed.

Mainline The piping upstream of the control valve.

Management Allowable Depletion (MAD) The percentage of water at field capacity which can be removed from the soil by the plant prior to reaching the wilting point, at which time irrigation should occur.

Master Valve A control valve installed at the beginning of the system which opens only during system operation.

Matched Application Rates When all sprinkler heads within a zone — regardless of the arc of coverage — have approximately the same application rates.

Net Positive Suction Head Required (NPSHR) The suction head or pressure required by the pump at the eye of the impeller to ensure that the liquid will not boil (cavitate) under the reduced pressure and the impeller will operate without cavitation. This information is available from the manufacturer and is usually indicated on the pump curve. It varies with different flow conditions.

Net Positive Suction Head Available (NPSHA) The suction head or pressure that is available at the eye of the impeller after head losses, elevation differences and vapor pressure are subtracted from the site atmospheric pressure. NPSHA should always exceed NPSHR by at least 3 feet of head to avoid cavitation, pump damage and loss of prime.

Nozzles A short duct that is used to direct the flow of water from sprinkler heads.

Operating Pressure The pressure at any point in the irrigation system when the system is functioning. Measured with flow in the line.

Percolation The slow passage of water downward through the soil.

Polyvinyl Chloride (PVC) A type of plastic used in the manufacturer of pipe, fittings and other system equipment. Semi-rigid pipe, usually white in appearance.

Polyethylene (PE) A type of plastic used in the manufacturer of pipe, fittings and other system equipment. Flexible pipe, usually black in appearance.

Potable Water Water suitable for drinking.

Precipitation Rate The quantity of rainfall deposited on the Earth and measured in inches per hour. Also used in lieu of application rate.

Pressure (PSI) The force that moves the water through the system. Pressure is measured in pounds per square foot or feet of head. One foot of water depth equals 0.433 PSI and one PSI equals 2.31 feet of head.

Pressure Gauge A gauge used for indicating water pressure.

Pressure Head The operating pressure in feet of head required at the sprinkler head.

Pressure Line See mainline.

Pressure Loss See friction loss.

Pressure Reducing Valve A device used to reduce water pressure.

Pressure Vacuum Breaker See backflow preventer.

Quick Coupler Valve A device attached to the system mainline which is activated by inserting a key attached to a sprinkler head or hose. Used for supplemental water.

Rain Shut-off Device Accumulates rainfall and interrupts the irrigation cycle once rainfall exceeds a predetermined amount.

Reduced Pressure Backflow Preventer See backflow preventer.

Repeat Cycles See cycles.

Riser An upright piece of pipe used to support nozzles, heads, emitters, backflow preventers and valves.

Root Zone The depth of the plant root zone in the soil.

Scheduling Coefficient A method of uniformity which relates the lowest precipitation rate in an area to the average precipitation rate — without regard to the size of the area — as measured in the DU calculation. A Scheduling Coefficient value of 2 indicates the driest area is two times as dry as the average of the entire area.

Shut-off Head The maximum head the pump will produce at a zero flow rate.

Sleeve A piece of pipe installed under paved areas to accommodate irrigation system pipe and wiring. It is used for ease of replacing equipment.

Soil Moisture Sensor A device used to monitor the amount of water in the soil.

Solenoid The electrical device on a control valve that actuates a plunger to open the valve.

Solvent Welded A chemical substance used to dissolve the outer surface of PVC pipe and the inside of fittings, blending them as one.

Spacing Distance between sprinkler heads.

Sprinkler Head A device used to spray water on lawns, landscapes and ornamentals.

Static Discharge head The elevation difference between the impeller eye and the sprinkler orifice.

Static Pressure An indication of the amount of force available to operate the irrigation system. Measured with no flow in the line.

Station A position in the controller that activates a certain zone.

Subsurface Drip Irrigation Usually pipe with drip emitters installed at 12- to 24-inch intervals inside the pipe or in the wall of the pipe. These pipes are installed 12- to 24-inches apart, at 3- to 4-inch depths.

Suction Head Occurs when the water level is *higher* than the pump impeller eye. This produces pressure from the weight of the water (measured in feet of head) against the impeller eye, which is added to the atmospheric pressure. This condition is referred to as a flooded suction or a positive static suction head.

Suction Lift Occurs when the water level is *lower* than the pump impeller eye. When the pump is located above the water source, a partial vacuum is created at the eye of the impeller. The difference between the atmospheric pressure on the water and the vacuum at the eye of the impeller drives water up the suction piping.

Normal atmospheric pressure at sea level is 14.7 PSI or 34 feet of head, but it's impossible to lift water 34 feet, even with a perfect vacuum at the impeller. Suction lift of up to 26 feet can be attained at sea level with careful planning. To be safe, suction lift should never exceed 15 feet at sea level or 12 feet at 5,000 foot elevations. The vertical distance the water is raised by the pump is also called static suction lift.

Surge Waves of high and low water pressure moving through the pipes. See water hammer.

Swing Joint Fittings between the pipe and head that are used to absorb surface impact of the head and prevent breakage.

Tensiometers A device to determine the water content of the soil by measuring the tension or vacuum as water is removed from the soil.

Transpiration The process of water moving through the leaf surface and vaporizing. It's used by the plant to cool its leaves.

Total Head The sum of all heads that affects the ability of the pumps to produce pressure at a given flow rate.

Total Static Head The difference in elevation between the water source and the point of discharge.

Uniformity High uniformity occurs when the same amount of water is applied to each square foot of turf or landscape. The uniformity of rainfall is 96 percent without wind. Fifty percent uniformity is common in older systems.

Vapor pressure The pressure at which water will change to vapor (boil) at a certain temperature. The vapor pressure of water increases with temperature and reduces the available pressure at the pump suction.

Velocity The rate of water flow calculated in feet per second (FPS).

Velocity Head The energy required to attain a certain velocity. The higher the velocity the greater the velocity head. Velocity head is usually insignificant in most landscape irrigation pumping applications, but should be calculated where low pressure and high velocities occur.

Water Hammer A series of water surges in pipes caused by a fast closing valve. The magnitude and frequency is dependent on the velocity of flow, size, length and material of the pipe.

Wire Gauge A standard method of sizing wire diameter. The smaller the number the larger the diameter of the wire. #12, #14 and #16 gauge wire is commonly used in the irrigation industry.

Wilting Point When the plant can no longer pull water from the soil.

Zone An area covered by a group of heads connected to a control valve.

ANSWERS

1. Sandy soil.
2. 0.10 inches per hour.
3. 0.75 inches of water.
4. 0.166 inches of water.
5. Apply 0.4 inches of water every other day.
6. Five minutes.
7. June in the Northern Hemisphere and December in the Southern Hemisphere.
8. Management allowable depletion forces oxygen to enter the soil as the plant pulls the water out.
9. Soil type and crop cover as well as compaction, slope and thatch will decrease the infiltration rate.
10. Soil preparation.

1. After 10 p.m. and before 6 a.m.
2. 0.55 inches per hour.
3. 0.40 inches per hour.
4. 231 cubic inches.
5. 471 minutes.
6. 63 minutes every fourth day.
7. The conversion of gallons to cubic inches of water, or inches per square foot.
8. 80 percent of ET.
9. It's time to irrigate.
10. The base times 0.866.

1. 50 feet.
2. 40 to 45 PSI, and the spacing has no effect on this.
3. 30 PSI.
4. Floating fine mist.
5. A Pitot tube and pressure gauge.
6. Yes, it increases proportionately to pressure increases.
7. Both situations cause a distortion of the spray pattern and a reduced radius of coverage resulting in poor uniformity of coverage.

8. Low water pressure.
9. By reducing the nozzle sizes if spacing is good. Then split the zone and add another electric control valve or install a booster pump.
10. Yes, especially if check valves are not installed in the heads and the lateral is drained after every cycle.

1. Yes.
2. 80 percent.
3. No, but they can improve efficiency.
4. Yes, using catchment devices.
5. Yes, because the operator will apply more water in order to avoid dry spots.
6. Five heads operating together.
7. Soils, irrigation scheduling, pressure control, uniformity and efficiency.
8. An Anemometer is a spinning device with cups mounted on a pole that measures wind speed.
9. Evaporation, wind drift, overspray, runoff and soil drainage.
10. The percentage of applied water that ends up in the plant root zone.

1. 200 mesh screen.
2. $\dfrac{261,360 \text{ sq. ft.} \times 0.6234 \times (1.6/0.6 = 2.66) = 433,398.6}{3 \quad \times \quad 6 \quad \times \quad 60 \quad = \quad 1,080}$ = 401.3 gpm + 30% = 521.69 gpm
3. Yes, whenever rust or scale is present in the supply.
4. Yes.
5. 0.0058 inches.
6. AWWA manual M-22.
7. Install an intake screen of filter to protect the pump, and additional filtration prior to the water entering the system.
8. Yes, from two to eight years.
9. The screen size should be 1/6th the size of the smallest orifice in the system.
10. Yes, a 30 percent safety factor is required for the variation in zone valve flow rate.

1. Equipment that is high in quality and performance, low in maintenance, long lasting and water conserving.
2. Small radius rotors.
3. Ease of operation and simplicity.
4. 25 to 30 PSI.
5. Check valves in the base of the head.
6. Rotor heads.

7. "Blow by" refers to the flushing action between the stem and body that occurs when the spray head pops up.

8. Flexible tubing and premanufactured PVC swing joints.

9. Install surge protection for the primary or 120-volt power source as well as on the secondary or 24-volt lines to the electric control valves.

1. Yes.

2. Triangular, square and rectangular.

3. In relatively flat planting beds.

4. 23 percent resulting in a spacing of 11.55 feet across the slope.

5. Sprinkler head selection and placement.

6. 111.8 feet.

7. Slope tops, low overhead clearances such as mature trees or bridges and in high wind areas.

8. Determine the available operating pressure range for sprinkler heads.

9. If the system static pressure is above 75 PSI with little elevation change.

10. Spray diffusion will distort the sprinkler pattern and lower uniformity.

1. 4:1.

2. 15 percent.

3. 0.433 PSI.

4. Yes.

5. The center of the zone.

6. No.

7. The top should be watered with a separate zone.

8. No, cheat on the maximum safe flow and it may come back to haunt you in high water costs, additional liability and water waste.

9. No, this can delay by minutes the time it takes for the first and last head on the line to pop-up. Center feeding the line will also reduce surge potential.

10. Safe maximum flow rate, plant water requirements and slope conditions.

1. Yes.

2. No.

3. 62.8 PSI.

4. No.

5. No.

6. Surge pressure is a series of pressure pulsations of varying magnitude, above and below the normal pressure in the pipe.

7. Each pipe model is categorized by the type of material and given a "C" value which indicates its relative smoothness or roughness.

8. 2.58 PSI.

9. 4.29 fps.

10. Yes, actual closure time ranges from 0.5 to 0.8 seconds.

1. 1 inch and smaller.

2. The American Water Works Association, Manual of Water Supply Practice, Sizing Water Service Lines and Meters, AWWA M22.

3. A maximum of 7.5 fps.

4. This reduces the cost of the valve with a relatively small increase in pressure loss.

5. Whenever the static site pressure is above 70 PSI.

6. Looping the mainline usually reduces construction costs and provides for a better system.

7. Size the meter at 50 percent or less of the AWWA maximum safe capacity for irrigation demand, and schedule irrigation operation during the night hours.

8. Fitting losses are calculated using equivalent pipe lengths from the table "Pressure Loss in Pipe Fittings and Valves" found in the Appendix.

9. The industry standard for pressure variation within the lateral line is 20 percent of the highest pressure in the lateral.

10. Always check the zone furthest from the water source and the zone with the highest flow.

1. Upstream.

2. 3 to 6 inches inside the edge of the root ball.

3. Drip distribution tubing should be installed with a minimum 4 to 6 inches of soil cover.

4. No, check the manufacturer's catalog for information.

5. Pressure compensating emitters

6. Filtration of any water source is required to ensure the longevity and uniformity of the drip system.

7. Self-cleaning emitters and excellent water filtration.

8. Yes.

9. Keep the pressure fluctuation within the zone under 10 percent.

10. For cost containment and ease of operation.

1. #10.

2. The force applied to the flow of electrical current.

3. 2 percent of the rated voltage.

4. Current is the flow of electricity.

5. In areas where a high incidence of lightning occurs.

6. The amperage required to actuate the solenoid or electrical device.

7. Yes.

8. 20 percent.

9. 1.61 volts.

1. The fabric keeps the soil from working its way up through the gravel.

2. Light compaction similar to adjacent soil.

3. No, sleeve should be 4 inches to 6 inches in diameter.

4. With a minimum straight pipe length equal to 10 times the nominal diameter of the pipe on the upstream side and up to 5 times the diameter on the downstream side.

5. Air valve or pressure gauge.

6. No, use a cable laying device or install in a trench.

7. The high humidity that develops in vaults can cause corrosion and greatly reduce the life span of the electrical equipment.

8. An air valve should be installed downstream of every pressure reducing valve to aid in adjusting the PRV and checking operating pressure.

9. Controller charts are reduced size, as-built drawings containing zone numbers with the zone coverage areas highlighted in different colors, laminated in 20-mil plastic and mounted in the controller door.

1. Mower damage to heads and valve boxes due to improper installation.

2. Compaction similiar to the undisturbed adjacent soil.

3. The top of the head should be about 1/8-inch below finished turf grade.

4. Yes, the surge potential is greater in systems without check valves in the heads because water will drain out of the lateral after every operation.

5. Threaded connections.

6. Pulling pipe is acceptable if the soil is free of sharp rocks or other conditions that may damage the pipe.

7. Allow at least 2 inches, or three finger widths, between the heads and the edge of a driveway or sidewalk where turf edging equipment will be used. Allow 6 inches in planting beds.

8. Always hand tamp soil firmly around the head to prevent movement and erosion and, when possible, install the head against undisturbed soil for greater stability.

9. Install the clamp over the serrated edges of the insert fitting.

10. A 14-inch to 24-inch length of highly resilient pipe that will not kink.

1. In backflow prevention devices and PVC pipe.

2. The frequency is determined by the quality of the equipment and installation.

3. Yes, a freeze/thaw cycle may occur during the winter.

4. Polyethylene NSF pipe.

5. Every three to four weeks.

6. Yes, from hot air.

7. Arc and radius coverage adjustments, proper head alignment, damaged equipment, potential liability, leakage, cleaning system filters, seeping control valves and slow closing valves.

8. After every mowing.

1. Use a non-abrasive device such as a plastic stir stick or plastic toothpick.

2. Yes, rust and other particles can plug heads and valve ports.

3. Suspended solids in the water can cause abrasive action which will enlarge the nozzle.

4. Match the performance to the previous head or to existing zone/head spacing and precipitation rates.

5. A bad solenoid or lack of electrical power from the controller.

6. Yes, if the turf height and soil buildup blocks the trajectory of spray.

7. To clean, remove the valve bonnet and use a small copper wire to clean the ports in the bonnet and body of the valve.

8. Seeping valves are usually caused by particles imbedded in the rubber seat of the diaphragm or lodged between the diaphragm and valve body. Additionally, the diaphragm may stretch out or develop cracks or a weak spring may be the culprit.

9. The pipe may be defective and/or the pressure could be too high.

10. Repairing or replacing equipment usualy depends on the cost of repair parts and labor vs. the cost of installing new equipment.

1. Install schrader valves downstream of the pressure reducing valve and each zone control valve.

2. A plugged nozzle or pipe, a defective pressure reducing valve or increased presssure from the water purveyor.

3. Yes, using an ohmmeter.

4. Using a ground fault locator.

5. A hot room will reduce the motor life by 50 percent for every 18 degrees in temperature above the motor nameplate rate.

6. A bearing failure or a misaligned drive coupling or connecting pipe.

7. Annually.

8. Changes in water pressure may occur in potable water supplies and pump systems because of increased area demand, equipment deterioration and/or pressure changes made by water purveyors.

9. Probably a broken common wire or a cut in all of the hot wires.

10. Test the resistance between each wire and an earth ground. If the reading is under 100 ohms, then a ground fault is indicated.

1. Because debris such as leaves can enter the sensing area and retain moisture resulting in false interruptions and dry lawns.
2. Flow sensors are used in systems to calculate water application, to identify excess flows or leaks and to log total water consumption.
3. Install a straight length of pipe on both the inlet and the outlet of the sensor to reduce turbulence.
4. ET gauges are modified Anemometers that simulate the evapotranspiration rate of turf and other crops.
5. Tensiometers, solid-state tensiometers, gypsum blocks and electrical resistance blocks.
6. They measure the matrix potential or capillary tension in the soil which is similar to the force a roost must exert to take water from the soil.
7. To prevent the chance of a liability claim from ice remaining on streeets and walks in colder climates.
8. Install a staight length of pipe (the same size as a sensor input and output) that is 10 pipe-size diameters long on the intake side of the sensor and another length of pipe on the outlet side that is six pipe-size diameters long.

1. 15 fps.
2. Low water pressure.
3. PRV or pressure regulating valve.
4. The cost of water and maintenance.
5 High water pressure.
6. Poorly designed irrigation systems that over spray.

20

1. To determine soil conditions and the potential for vandalism.
2. Actual job costing information.
3. Add the columns both down and across and have someone else check the bid.
4. Payroll taxes, workmen's compensation insurance, medical insurance, retirement benefits, overtime, holiday, vacation and sick pay.
5. Determine if the project fits his business and can be profitable.
6. Yes, if payments take 45 to 60 days and some money is retained until acceptance.
7. The prime contractor position gives the firm more control over scheduling, payments, negotiations and disputes.

1. Sites that are more than 12 years old or sites that were poorly designed and installed.
2. An extensive background in irrigation system design, installation and maintenance with certification as an irrigation designer (CID), landscape irrigation auditor (CLIA) and irrigation manager (CIM).
3. Historical water usage analysis.
4. Water usage is measured in units that are either in thousands of gallons (M gallons) or per 100 cubic feet (CCF).
5. Annual water use divided by irrigated area times 12 equals annual inches.
6. They can operate after the next valve starts its cycle resulting in low pressure for both zones and inadequate coverage.
7. 38.77 inches annually (1,850,000/60,000 = 30.83/0.6234 = 49.46 inches).

1. They are usually repairs for leaks or hazards and minor equipment adjustments.
2. Describe the equipment by brand name, model, size, age, maintainability, condition, etc.
3. They are compared to manufacturers' recommended optimum pressures.
4. It is a rating system of irrigation components which indicates whether repair or replacement is the most cost effective method.
5. Condition, pressure, efficiency and conservation.

23

1. Suction lift occurs when the water level is lower than the pump impeller eye.
2. H_a = 32.8 feet of head at 2,000 feet, derated 85 percent 27.88 feet

   H_s = suction lift   -9.0 feet

   H_f = friction losses   -2.85 feet

   H_{vp} = vapor pressure   -0.839 feet

   NPSHA 15.191 feet

3. 1.171 feet of head or 0.507 PSI.
4. 0.314 velocity head.
5. Shutoff head is the maximum head the pump will produce at a zero flow rate.
6. 1.007.
7. 58 gpm.

INDEX

Application rate 15, 18, 19, 21, 22, 33, 34, 35, 36, 57, 58, 61, 65, 67, 155
Arc of coverage 58, 116
Atmospheric pressure 69, 159
Available water holding capacity 12

Backflow preventer 28, 42, 43, 67, 78, 99, 100, 109, 111, 112, 113, 121, 122, 123, 131, 135,
    143, 145, 154

Cavitation 162
Check valve 28, 49, 57, 104, 112, 118, 152, 154, 165
Control valve 28, 41, 49, 50, 63, 66, 67, 73, 78, 79, 85, 86, 91, 97, 98, 99, 105, 106, 109, 116,
    118, 121, 122, 123, 124, 128, 129, 130, 131,136, 143, 151
Controller 21, 22, 23, 34, 37, 50, 51, 52, 92, 93, 94, 95, 99, 100, 111, 113, 117, 123, 124, 129,
    137, 143, 145, 150, 153, 156, 157
Coverage 19, 25, 26, 27, 29, 31, 32, 33, 34, 36, 49, 56, 57, 58, 59, 61, 67, 75, 76, 79, 87, 100,
    106, 109, 116, 134, 136, 137, 151, 154, 157
Crop Coefficient 20, 21
Cycle 15, 22, 28, 47, 49, 51, 67, 85, 110, 113, 136, 151, 155

Design 29, 31, 37, 39, 45, 47, 55, 56, 58, 61, 63, 65, 69, 72, 75, 76, 77, 78, 84, 86, 92, 97, 99,
    121, 129, 133, 134, 135, 137, 138, 148, 155, 156, 157, 160,
Distribution uniformity 33
Drain valve 111, 112, 113, 143

M

N

O

P

Q

Rain shut-off device 37, 101

Reduced pressure backflow preventer 154

Repeat cycles 19, 22, 51, 57, 137

Riser 73, 87, 118, 131

Root zone 12, 13, 15, 20, 21, 31, 36, 83, 87, 88, 128, 131, 137

Sleeve 99

Soil 11, 12, 13, 14, 15, 19, 20, 21, 22, 25, 31, 37, 38, 51, 55, 57, 63, 67, 83, 84, 85, 86, 87, 88, 94, 95, 97, 98, 100, 104, 105, 106, 111, 115, 116, 117

Soil moisture sensor 22, 37, 128, 129, 130, 131

Solar radiation 14, 21, 128

Solenoid 51, 91, 92, 94, 95, 98, 117, 123, 124, 138

Solvent welded 104

Spacing 11, 19, 28, 31, 32, 37, 58, 59, 60, 61, 116, 134, 137, 154, 157

Sprinkler head 11, 15, 18, 19, 25, 28, 32, 33, 39, 42, 45, 51, 56, 58, 59, 63, 76, 91, 104, 112, 121, 122, 136, 137

Static pressure 25, 28, 29, 42, 45, 56, 57, 77, 78, 122, 136

Station 14, 20, 22, 113, 121, 123, 124, 128, 155

Suction head 41, 160, 165

Suction lift 161, 164

Surge 22, 49, 50, 66, 72, 73, 78, 85, 94, 95, 104, 105, 111, 112

Swing joint 51, 104, 105, 137

Tensiometers 129

Total head 161

Transpiration 11, 14

Uniformity 25, 29, 31, 32, 33, 34, 35, 36, 37, 43, 44, 45, 48, 49, 58, 59, 61, 67, 79, 85, 106, 115, 116, 134, 137, 151, 154

NOTES

NOTES

NOTES

MAIN DISHES
FOR EVERY OCCASION

MAIN DISHES

FOR EVERY OCCASION

Rosemary Wadey

CONTENTS

ANOTHER BEST-SELLING VOLUME FROM HPBooks®

Publisher: Rick Bailey; Editorial Director: Retha M. Davis
Editor: Jeanette P. Egan; Art Director: Don Burton
Book Assembly: Leslie Sinclair
Book Manufacture: Anthony B. Narducci
Typography: Cindy Coatsworth, Michelle Claridge
Recipe testing by International Cookbook Services: Barbara Bloch,
President; Rita Barrett, Director of Testing

Published by HPBooks, Inc.
P.O. Box 5367, Tucson, AZ 85703   602/888-2150
ISBN 0-89586-337-5
Library of Congress Catalog Card Number 84-81922
© 1985 HPBooks, Inc.   Printed in the U.S.A.
1st Printing

Originally published as Special Occasion Casseroles
© 1983 Hennerwood Publications Limited

Cover Photo: Shrimp Veracruz-Style, page 60
Photo and recipe first appeared in *Mexican Cookery* by Barbara Hansen, HPBooks

Introduction

A main dish is the center of any meal, but is even more important to a special-occasion meal. A good example is stuffed turkey at Thanksgiving or glazed ham at Easter. When a host or hostess starts to plan a meal, the main dish is generally considered first. Then side dishes are selected that will complement and highlight the main dish without overshadowing it. Remember to contrast shapes, colors, textures and temperatures in a meal for added interest. Main dishes that combine several ingredients may only need a crisp, green salad and a loaf of crunchy bread to make a complete meal.

Recipes in this book have been selected to fit various categories. Some are quick and easy, requiring only minutes to prepare and cook. Others simmer slowly for hours to blend flavors and tenderize meat. Economy is also considered. Several recipes use chicken and less-tender cuts of meat. Many recipes are both elegant and economical. Some recipes are included for more expensive meats for those extra-special meals. Many recipes include suggestions for accompaniments and garnishes.

Most recipes serve four, however many can be doubled for additional servings or to make an extra dish for freezing. Adjust cooking times, as needed, if recipes are doubled.

ADVANCE PREPARATION & FREEZING

Many main dishes freeze successfully, particularly stews and casseroles. Avoid freezing those with cream, egg yolks, hard-cooked eggs or potatoes. These ingredients don't freeze well. Most main dishes freeze well up to 2 to 3 months. If a dish is highly spiced, 2 to 3 weeks is long enough in the freezer. Frozen dishes can be reheated while still frozen, or first thawed in the refrigerator or microwave oven. Reheating must be done thoroughly; once the dish starts to bubble, bake 15 to 20 minutes. Food can be frozen in a foil-lined casserole. Remove food and foil once contents are frozen; wrap tightly. Meanwhile, the casserole dish can be used for other purposes. To reheat, place frozen food in original casserole dish. Other suitable freezer containers include plastic freezer containers, plastic freezer bags, freezer paper and heavy-duty foil. Remember to leave a 1-inch headspace in freezer containers to allow for food expansion.

Casseroles and stews are often more flavorful if made the day before serving. If a dish is prepared in advance, cook dish for all but the last 20 minutes. Refrigerate until needed. Before serving, bake at 325F (165C) or simmer on low heat 45 minutes to 1 hour.

SLOW COOKERS & PRESSURE COOKERS

Slow cookers and pressure cookers can be convenient and economical. The slow cooker, after preliminary preparation, cooks food on low heat, usually taking 6 to 8 hours or longer. The advantages are that once under way, it needs no attention, and the amount of electricity used is minimal. It is ideal for the busy housewife or business woman who likes to come home to a hot evening meal. Small-size slow cookers are great for one or two persons.

A pressure cooker reduces cooking time because the increased pressure lets food cook at higher than usual temperatures. Using a pressure cooker is an efficient way to quickly cook less-tender meat cuts. It can be a wonderful time-saver for the creative but busy cook.

With both slow cookers and pressure cookers, it is essential to read the manufacturers' instructions carefully before using. Always follow them while cooking. With minor adjustments, made according to the manufacturers' directions, many recipes can be adapted for use with either a slow cooker or pressure cooker.

MEAT

All types of meat are included—beef, lamb, pork, veal, ham and even oxtail. All can be turned into something special. Wines, sherry, Madeira, brandy and other types of alcohol are added to a number of recipes to enhance flavors and help tenderize meat. Many old favorites appear along with plenty of new ideas.

POULTRY & GAME

This chapter presents a selection of exciting recipes suitable for family and friends. There is a wide variety of recipes using chicken, turkey, duck and Cornish hens. Many are fast and easy to prepare. Poultry is an ideal choice for an affordable and delicious main dish.

FISH

Fish and shellfish are becoming more widely available throughout the country. Rapid transportation and a wide selection of frozen products mean that you no longer have to live by a seacoast to enjoy good seafood. Fish and shellfish are excellent choices for festive occasions because they can be prepared quickly. A good rule to follow is to cook fish 10 minutes per inch of thickness. Do not overcook fish or it may become dry and tasteless.

FAVORITES FROM HOME & ABROAD

As an interesting contrast to the other chapters, recipes included here are a mixture of special ideas. Here the French classics appear alongside dishes from Poland, Australia, New Zealand, Switzerland and the United States. Ingredients are often well-known, but used in surprising combinations.

COOKING TERMS

Beurre manié: Paste made by blending flour and room-temperature butter. Small amounts of beurre manié can be beaten into hot liquids until desired thickness is obtained. The usual combination is 2 parts butter to 3 parts flour; exact proportions will vary.

Bouquet garni: Term used for an herb package containing 1 bay leaf, 2 parsley sprigs and thyme. If the herbs are fresh, they can serve as their own package. If dried herbs are used, they are usually wrapped in cheesecloth to make removal easy after the dish is cooked. Other ingredients, such as peppercorns or celery leaves, can be added, if desired.

Deglazing: Technique in which browned pan drippings are loosened and incorporated into sauce by heating the roasting pan, then stirring in wine, stock or other liquid. This adds a rich flavor and color to sauce.

Degrease: A method of removing fat from the surface of casseroles, pan drippings, stocks and stews by skimming the surface with a spoon or bulb baster or by chilling until fat hardens on the surface and can be lifted off. More complete removal of fat results from the chilling method. This method is particularly good for stews, soups and casseroles that improve in flavor when made in advance.

Stock: Stock is made by simmering meat, chicken or fish bones and trimmings with vegetables, herbs and water. Stock must be strained and degreased before using. Homemade stock can be frozen. If homemade stock is not available, use canned bouillon or broth. Bouillon cubes or granules are also available. Remember that prepared products may be higher in salt; adjust seasoning accordingly to avoid too salty a flavor.

Beef Burgundy

2 tablespoons vegetable oil
1-1/2 lbs. beef-round steak, cut into 2" x 1" strips
2 onions, sliced
1 garlic clove, crushed
1-1/2 cups red wine
1 tablespoon tomato paste
1 bay leaf
2 teaspoons Worcestershire sauce
2/3 cup pitted prunes
Salt
Freshly ground pepper
1 tablespoon cornstarch
Water
4 oz. button mushrooms, halved

To garnish:
Grilled bacon rolls, see below
Parsley sprigs, if desired

1. Preheat oven to 325F (165C).
2. Heat oil in a large skillet over medium heat. Add beef strips; sauté until browned. Add onions and garlic; continue cooking a few minutes, stirring frequently. Transfer to a 3-quart casserole; set aside.
3. Add wine to skillet; bring to a boil. Stir in tomato paste, bay leaf, Worcestershire sauce and prunes. Season with salt and pepper. Pour wine mixture over browned beef strips. Cover casserole with foil or lid.
4. Bake in preheated oven 2 hours.
5. In a small bowl, blend cornstarch with a little cold water. Stir cornstarch mixture into hot casserole along with mushrooms and some water, if necessary. Discard bay leaf. Bake 15 minutes longer or until mushrooms are cooked. Serve hot, garnished with bacon rolls and parsley, if desired. Makes 4 servings.

To make *Bacon Rolls*, stretch bacon slices evenly on a board using the back of a knife. This will make bacon more even and easier to roll up. If you are worried about rolls unwinding during cooking, thread several on a metal skewer or push a wooden pick into each roll. Broil until crispy.

Beef Carbonnade

3 tablespoons all-purpose flour
Salt
Freshly ground pepper
1-1/2 lbs. beef stew cubes
1/4 cup vegetable oil
2 large onions, thinly sliced
1 to 2 garlic cloves, crushed
1 cup dark beer or ale
1 cup beef stock
2 tablespoons ketchup or 1 tablespoon tomato paste
Pinch of ground mace or nutmeg
1 bay leaf
2 teaspoons brown sugar
2 teaspoons vinegar
1-1/2 teaspoons prepared brown mustard
3 to 4 carrots, cut into 3- to 4-inch sticks
4 oz. button mushrooms

To garnish:
Chopped fresh parsley, if desired

1. Preheat oven to 325F (165C).
2. In a plastic bag, combine flour, salt and pepper. Add beef cubes; shake to coat. Heat 3 tablespoons oil in a large skillet over medium heat. Add seasoned beef cubes; sauté until browned. With a slotted spoon, transfer to a 3-quart casserole.
3. In same skillet, heat remaining oil. Add onions and garlic; sauté until lightly colored. Stir in remaining seasoned flour; cook 1 minute, stirring constantly.
4. Gradually stir in beer or ale and stock; bring to a boil, stirring frequently. Stir in ketchup or tomato paste, mace or nutmeg, bay leaf, sugar, vinegar and mustard. Season to taste with salt and pepper. Pour beer mixture over browned beef cubes.
5. Add carrot sticks to casserole; stir well. Cover tightly with foil or lid; bake in preheated oven 1-1/4 hours.
6. Add mushrooms to hot casserole. Bake 25 to 30 minutes longer or until beef is tender. Discard bay leaf. Serve sprinkled with chopped parsley, if desired. Makes 4 servings.

Top to bottom: Beef Burgundy, Beef Carbonnade

Boeuf à l'Orange

2 oranges
3 tablespoons vegetable oil
1-1/2 lbs. beef stew cubes
8 oz. small white onions
1 garlic clove, crushed
2 tablespoons all-purpose flour
1 cup beef stock
1 tablespoon tomato paste
3 tablespoons brandy
1 tablespoon molasses
Salt
Freshly ground pepper
4 oz. mushrooms, thickly sliced

To garnish:
Parsley sprigs
Orange wedges or slices

1. Preheat oven to 325F (165C).
2. With a vegetable peeler, remove peel from oranges; cut peel into julienne strips. Juice both oranges. Set julienned peel and juice aside.
3. Heat oil in a large skillet over medium heat. Add beef cubes; sauté until browned. With a slotted spoon, transfer to a 3-quart casserole.
4. Sauté onions and garlic in same fat until golden brown. With a slotted spoon, transfer to casserole.
5. Stir flour into fat in skillet; cook 1 minute, stirring constantly. Gradually stir in stock; bring to a boil. Add orange juice and strips of peel.
6. Stir tomato paste, brandy and molasses into sauce. Season with salt and pepper. Pour sauce over browned beef cubes. Cover casserole tightly with foil or lid.
7. Bake in preheated oven 2 hours.
8. Add mushrooms to hot casserole. Add extra stock, if necessary. Bake 30 minutes longer or until beef is tender.
9. To serve, garnish with parsley sprigs and orange wedges or slices. Makes 4 servings.

Clockwise from bottom: Fillet of Beef Dijon, Boeuf à l'Orange, Swiss Steak

Swiss Steak

2 lbs. beef-round steak, cut into 8 serving pieces
Salt
Freshly ground pepper
2 tablespoons vegetable oil
1 (16-oz.) can tomatoes
2 onions, sliced
1 garlic clove, crushed
2 tablespoons all-purpose flour
1/2 cup red wine
1/2 cup beef stock
1 tablespoon tomato paste
2 tablespoons capers
1 tablespoon wine vinegar

To garnish:
Parsley sprigs, if desired

---

1. Preheat oven to 350F (175C).
2. Sprinkle beef with salt and pepper. Heat 1 tablespoon oil in a large skillet over medium heat. Add beef; sauté until browned. Transfer to a 3-quart casserole.
3. Drain tomatoes, reserving juice. Add drained tomatoes to casserole with beef. Add remaining oil to skillet. Add onions and garlic; sauté until lightly browned.
4. Stir in flour; cook 1 minute, stirring constantly. Stir in wine, stock and reserved tomato juice. Boil mixture 2 minutes, stirring frequently; add tomato paste, capers and vinegar. Season mixture with salt and pepper; pour over browned beef.
5. Cover casserole with foil or lid; bake in preheated oven 2 to 2-1/2 hours or until beef is tender.
6. Serve garnished with parsley, if desired. Makes 4 to 6 servings.

Variation
Veal stew cubes make a good alternative to beef in this recipe. To change the flavor of this casserole, omit capers and add 4 ounces dried apricots. Alternatively, put prepared ingredients into a shallow casserole; cover with a layer of thinly sliced potatoes—about 1-1/2 pounds. Brush potatoes with melted butter; bake uncovered about 2 hours or until meat is tender, and potatoes are browned and crispy.

Fillet of Beef Dijon

1 (1-1/2 to 1-3/4 lb.) beef-loin tenderloin roast
Salt
Freshly ground pepper
2 tablespoons butter or margarine
Juice of 1 orange
Juice of 1 lemon
1/4 cup beef stock
1 to 2 tablespoons wine vinegar
1 tablespoon Dijon-style mustard
2 tablespoons brandy
6 tablespoons half and half
2 teaspoons cornstarch
1 tablespoon finely chopped small sweet
  pickles, if desired

To garnish:
Sautéed button mushrooms
Sautéed bread crescents
Parsley sprigs

Ask butcher for thick end of beef-loin tenderloin (rump end) and a thin piece of fat same size as roast to keep it moist during cooking. If you do not have the fat, rub roast with butter or margarine before roasting.

1. Preheat oven to 400F (205C).
2. Trim roast; season with salt and pepper. If using fat as suggested, tie it evenly around roast with string. Place roast in a shallow roasting pan just large enough to hold it.
3. Roast in preheated oven about 35 minutes or to desired doneness, basting once with pan drippings. For well-cooked beef, increase cooking time by 10 to 15 minutes.
4. Transfer roast to a serving plate; remove string and fat. Cover roast with foil to keep warm.
5. Remove excess fat from pan drippings. In a small bowl, combine fruit juices, stock and vinegar, according to taste; use to deglaze roasting pan.
6. Bring deglazing mixture to a boil. Stir in mustard and brandy. In a small bowl, combine half and half and cornstarch; stir into hot mixture. Bring mixture to a simmer; cook until thickened, stirring frequently. Season with salt and pepper; add pickles, if desired. Pour sauce into a gravy boat or small pitcher.
7. Garnish roast with mushrooms, bread crescents and parsley; accompany with sauce. Makes 4 servings.

Roast Beef with Ale

1 (3-lb.) beef-round rump roast
1-1/2 cups pale ale or beer
1 tablespoon vegetable oil
2 onions, sliced
1 garlic clove, crushed
Salt
Freshly ground pepper
2 bay leaves
1 tablespoon brown sugar
6 whole cloves
4 carrots, quartered
2 turnips, thickly sliced
1 tablespoon cornstarch
Water

To garnish:
Chopped fresh parsley

---

1. Preheat oven to 350F (175C).
2. Place roast in a glass dish just large enough to hold it; add ale or beer. Cover and refrigerate at least 24 hours (up to 48 hours); turn roast several times.
3. Drain roast, reserving marinade; pat dry with paper towels. Heat oil in a large skillet over medium heat. Add roast; sauté until well browned. Transfer to a large baking pan.
4. To fat remaining in skillet, add onions and garlic; sauté until lightly browned. Drain excess fat from skillet. Add reserved marinade; bring to a boil.
5. Add salt, pepper, bay leaves, sugar and cloves to hot marinade. Pour over roast. Cover pan tightly with foil or lid; bake in preheated oven 1-1/2 hours.
6. Add carrots and turnips to baking pan. Baste roast; replace cover. Bake about 1 hour longer or until beef is tender. Discard bay leaves.
7. Drain juices into a medium saucepan; skim off fat. In a small bowl, combine cornstarch with a little cold water. Stir cornstarch mixture into pan juices. Boil 1 minute or until thickened, stirring frequently. Season with salt and pepper. Pour sauce into a gravy boat or small pitcher.
8. Put beef on a warmed plate; surround by vegetables. Sprinkle with parsley. Serve sauce separately. Any leftovers are excellent served cold with salads. Makes 4 to 6 servings.

Variation

Two pounds of beef stew cubes can be used for this dish. Marinate beef cubes in ale as for the roast; follow above directions but bake 1-1/2 to 2 hours. After juices have been thickened, pour them back over beef and vegetables before serving. All dishes made with beer are even better if made the day before serving. Before serving, reheat about 1 hour.

Beef Rolls with Pecans

4 thin slices of beef-round steak, about 8" x 4"
Stuffing:
1/2 cup uncooked long-grain white rice
Salt
1-1/4 cups water
2 tablespoons vegetable oil
1 onion, chopped
1/2 teaspoon dried leaf thyme
1 tablespoon chopped fresh parsley or 1/2 teaspoon dried leaf parsley
1/3 cup chopped pecans
Freshly ground pepper
A little ground coriander
1 egg, lightly beaten

Sauce:
2 tablespoons vegetable oil
2 tablespoons all-purpose flour
1 cup beef stock
1/4 cup medium-dry sherry

To garnish:
1/4 cup dairy sour cream
A few pecan halves
Chopped fresh parsley

---

1. Preheat oven to 350F (175C). Pound steak slices between 2 sheets of plastic wrap or waxed paper until 1/4 inch thick.
2. For stuffing, in a medium saucepan over low heat, cook rice in 1-1/4 cups boiling salted water 12 to 14 minutes or until tender and water is absorbed.
3. Heat oil in a medium skillet over medium heat. Add onion; sauté until soft. Remove skillet from heat. Add thyme, parsley, pecans, salt, pepper, coriander and cooked rice. Cool rice mixture slightly; stir in beaten egg. Stuffing mixture should be fairly loose.
4. Divide stuffing among beaten beef slices. Roll each slice carefully to enclose stuffing. Secure each roll with wooden picks.
5. Heat oil in a large skillet over medium heat. Add beef rolls; sauté until browned. With tongs, transfer browned rolls to a shallow casserole or baking pan large enough to hold them in one layer.
6. Stir flour into juices left in skillet; cook 1 minute, stirring constantly. Gradually stir in stock and sherry. Bring sauce to a boil.
7. Season sauce with salt and pepper; pour over beef rolls. Cover pan with foil or lid. Bake in preheated oven about 1-1/4 hours or until beef is tender.
8. To serve, remove wooden picks from rolls; top each roll with a spoonful of sour cream, a few pecan halves and chopped parsley. Makes 4 servings.

Left to right: Beef Rolls with Pecans, Steak & Potato Dinner

Steak & Potato Dinner

2 tablespoons vegetable oil
1-1/2 lbs. beef-round steak, cut into 1-1/2" x 1" strips
2 large onions, sliced
1 (16-oz.) can tomatoes
8 oz. carrots, sliced
1 green pepper, sliced
2 tablespoons all-purpose flour
1 cup beef stock
1 tablespoon prepared brown mustard
1 tablespoon Worcestershire sauce
1 tablespoon soy sauce
Salt
Freshly ground black pepper
1-1/2 lbs. potatoes, peeled, sliced (about 4-1/2 cups)

1. Preheat oven to 350F (175C).
2. Heat oil in a large skillet over medium heat. Add beef strips; sauté until well browned. With a slotted spoon, remove beef strips from skillet; set aside.
3. Add onions to fat remaining in skillet; sauté until golden brown. Drain tomatoes, reserving juice.
4. In a 3-quart casserole, layer browned beef strips, sautéed onions, carrots, green-pepper slices and drained tomatoes.
5. Stir flour into fat in skillet; cook 1 minute, stirring constantly. Gradually stir in stock and reserved tomato juice. Bring mixture to a boil; stir in mustard, Worcestershire sauce and soy sauce. Season mixture with salt and pepper.
6. Pour hot mixture over contents of casserole. Top with sliced potatoes.
7. Cover casserole with foil or lid. Bake in preheated oven 1-1/2 hours.
8. Remove cover from casserole. Increase oven heat to 400F (205C). Bake 45 minutes longer or until potatoes are tender and golden brown. Makes 4 servings.

Deviled Meatballs

Meatballs:
1-1/4 lbs. lean ground beef
3/4 cup fresh breadcrumbs
1 small onion, finely chopped
Salt
Freshly ground pepper
1 tablespoon Worcestershire sauce
2 tablespoons vegetable oil

Sauce:
1 tablespoon all-purpose flour
1-1/2 teaspoons dry mustard
1-1/2 teaspoons Dijon-style mustard
1 tablespoon soy sauce
1 tablespoon Worcestershire sauce
1 tablespoon sweet chutney
1 cup beef stock
8 oz. carrots, cut into thin sticks
1 large cooking apple, peeled, cored, diced

To garnish:
Watercress or parsley

1. Preheat oven to 350F (175C).
2. In a medium bowl, combine ground beef, breadcrumbs, onion, salt, pepper and Worcestershire sauce. Shape into 16 equal balls.
3. Heat oil in a large skillet over medium heat. Add meatballs; sauté until browned. Remove from skillet; set aside. Pour off all but 1 tablespoon fat.
4. To make sauce, stir flour and dry mustard into fat remaining in skillet. Then stir in Dijon-style mustard, soy sauce, Worcestershire sauce, chutney and stock; bring mixture to a boil, stirring constantly. Season with salt and pepper.
5. Lay carrots and apples in a 2-1/2-quart casserole; arrange meatballs on top of carrots and apples. Pour sauce over meatballs. Cover casserole with foil or lid. Bake in preheated oven 45 minutes or until meatballs are done and carrots are tender.
6. To serve, remove any fat from surface of sauce. Stir casserole lightly; garnish with watercress or parsley. Spaghetti or noodles make a good accompaniment. Makes 4 servings.

Left to right: Deviled Meatballs, Spicy Oven Beef

Curried Beef with Pineapple

2 tablespoons vegetable oil
1-1/2 lbs. beef stew cubes
1 (8-oz.) can crushed pineapple, juice pack
3 large onions
2 teaspoons curry powder
1 tablespoon tomato paste
1 (16-oz.) can tomatoes
1 tablespoon wine vinegar
2 tablespoons apricot jam
1/2 cup beef stock
Salt
Freshly ground pepper

To garnish:
Chopped fresh parsley

This recipe has a mild curry flavor, suitable even for those who dislike curry. Lamb, pork or any kind of poultry can be used instead of beef. If using lamb or pork, cut cooking time to 1-1/2 to 1-3/4 hours. Turkey, cut into cubes, is also suitable and needs only 1 to 1-1/4 hours cooking. This dish is also good served with rice cooked with peppers, chopped toasted nuts, raisins and peas.

1. Preheat oven to 325F (165C).
2. Heat oil in a large skillet over medium heat. Add beef cubes; sauté until evenly browned. Transfer to a 3-quart casserole.
3. Drain pineapple, reserving juice. Chop onions very finely, preferably in a food processor. Add chopped onions and pineapple to skillet along with remaining ingredients, including reserved pineapple juice. Bring mixture to a boil. Pour hot pineapple mixture over beef cubes; stir well. Cover casserole tightly with foil or lid.
4. Bake in preheated oven 2-1/2 hours or until beef is tender. Season to taste.
5. Garnish with chopped parsley. Serve with hot cooked rice or creamed potatoes. Makes 4 servings.

Spicy Oven Beef

1-1/2 lbs. beef-round steak, cut into 4 serving pieces
Salt
Freshly ground pepper
1 teaspoon ground coriander
1 teaspoon ground ginger
2 tablespoons vegetable oil
16 small white onions
3 celery stalks, cut into 1-inch pieces
1/4 cup all-purpose flour
2 cups beef stock
1 tablespoon Worcestershire sauce
2 tablespoons whipping cream
3 to 4 tablespoons prepared horseradish

To garnish:
Celery leaves

---

1. Preheat oven to 325F (165C).
2. Sprinkle beef with salt, pepper, coriander and ginger; rub salt and spices into each piece. Heat oil in a large skillet over medium heat. Add beef; sauté until browned. Use a slotted spoon to transfer browned beef to a large casserole or baking pan.
3. Add onions and celery to fat remaining in skillet; sauté until soft. Stir in flour; cook 1 minute, stirring constantly. Gradually stir in stock. Bring mixture to a boil; add salt, pepper and Worcestershire sauce. Pour stock mixture over beef.
4. Cover casserole with foil or lid. Bake in preheated oven 2 hours or until beef is almost tender.
5. Stir in cream and horseradish. Bake 30 minutes longer. Garnish with celery leaves. Makes 4 servings.

Beef-Noodle Casserole

6 oz. green noodles
1-1/2 lbs. lean ground beef
2 onions, chopped
1 to 2 garlic cloves, crushed
2 teaspoons cornstarch
1 (16-oz.) can tomatoes
1/2 cup beef stock
1 tablespoon soy sauce
1 tablespoon Worcestershire sauce
1 tablespoon tomato paste
1 teaspoon dried leaf oregano
Salt
Freshly ground pepper
2 tablespoons butter or margarine
1/4 cup all-purpose flour
1 cup milk
1/2 cup grated sharp Cheddar cheese (1-1/2 oz.)

---

1. Preheat oven to 400F (205C).
2. Cook noodles according to package directions until almost tender; drain noodles. Grease a shallow baking dish; set aside.
3. In a large skillet over medium heat, sauté ground beef with no extra fat until browned. Add onions and garlic; sauté 3 to 4 minutes.
4. In a small bowl, blend cornstarch with some juice from tomatoes. Stir cornstarch mixture into browned beef. Add tomatoes with rest of juice, stock, soy sauce, Worcestershire sauce, tomato paste, oregano, salt and pepper. Boil 2 minutes, stirring frequently.
5. Put half the noodles in greased baking dish. Cover with beef mixture; add remaining noodles.
6. Melt butter or margarine in a medium saucepan. Stir in flour; cook 1 minute. Gradually stir in milk. Boil 1 minute, stirring constantly. Season to taste. Pour over noodles.
7. Sprinkle cheese over noodles. Cover casserole with foil or lid. Bake in preheated oven 15 minutes. Uncover casserole; bake 10 to 15 minutes longer or until cheese topping is brown and crispy. Makes 4 to 5 servings.

Variation
Four ounces of sliced mushrooms may be sautéed with the beef. Egg noodles or whole-wheat noodles may be used in place of green noodles.

Gingered Beef

1/2 cup all-purpose flour
Salt
Freshly ground pepper
1-1/2 to 2 teaspoons ground ginger
1-1/2 lbs. beef stew cubes
3 tablespoons vegetable oil
1 tablespoon grated fresh gingerroot
1 garlic clove, crushed
2 onions, sliced
1 cup beef stock
1 cup canned tomatoes
2 tablespoons vinegar
1 tablespoon honey
1 tablespoon Worcestershire sauce
1 (15-oz.) can red kidney beans or white beans, drained

This casserole has a rich tangy flavor which soon becomes a favorite, particularly because it is easy to prepare. It can be cooked in quantity and frozen up to 2 months—no longer because of its spiciness. For freezing, cook up to point of adding beans, but do not add them. Cool casserole rapidly; pack beef mixture into plastic containers or plastic freezer bags. To shape bags while filling, stand them in a bowl. Alternatively, line a casserole with foil. Add beef mixture; freeze until solid. Remove frozen beef mixture from casserole; wrap in foil and return to freezer. Remember, anything containing liquid will expand during freezing, so always allow at least 1 inch of headspace to prevent the container or bag from exploding during freezing. Thaw or partially thaw in refrigerator before cooking; add beans when ice melts. Cook 1 hour in a moderate oven (350F, 175C).

1. Preheat oven to 325F (165C).
2. In a plastic bag, combine flour, salt, pepper and ground ginger to taste. Add beef cubes; shake to coat.
3. Heat oil in a large skillet over medium heat. Add seasoned beef cubes; sauté until browned. With a slotted spoon, transfer to a 2-quart casserole. Add gingerroot, garlic and onions to fat remaining in skillet; sauté until lightly browned. Stir in remaining seasoned flour; cook 1 minute, stirring constantly.
4. Gradually stir in stock, tomatoes, vinegar, honey and Worcestershire sauce; bring mixture to a boil. Pour hot mixture over browned beef; cover casserole with foil or lid. Bake in preheated oven 1-3/4 hours or until beef is almost tender.
5. Add drained beans; stir through casserole. Cover and bake 25 to 30 minutes longer. Serve with boiled potatoes and salad. Makes 4 servings.

Biscuit-Topped Beef Bake

1-1/4 lbs. lean ground beef
1 large onion, sliced
2 carrots, diced
1-1/4 cups beef stock
1-1/2 tablespoons Angostura bitters
1 tablespoon tomato paste
Salt
Freshly ground pepper
2 teaspoons cornstarch
Water

Biscuit topping:
1-1/2 cups all-purpose flour
2 teaspoons baking powder
1/2 teaspoon salt
Freshly ground pepper
1/2 teaspoon Italian seasoning
1/4 cup butter or margarine
1/2 cup plus 2 tablespoons milk
Sesame seeds, if desired

1. Preheat oven to 350F (175C).
2. In a large skillet over medium heat, sauté beef with no extra fat, stirring constantly, until no longer pink. Add onion and carrots; cook 3 minutes. Drain off fat.
3. Add stock, bitters, tomato paste, salt and pepper to beef mixture; bring mixture to a boil. In a small bowl, blend cornstarch with a little cold water. Stir into beef mixture; cook until slightly thickened, stirring frequently.
4. Pour beef mixture into a 2-1/2-quart casserole. Cover with foil or lid. Bake in preheated oven 20 minutes.
5. To make biscuit topping, in a medium bowl, combine flour, baking powder, salt, pepper and Italian seasoning. Cut butter or margarine into dry ingredients until mixture resembles coarse crumbs. Add 1/2 cup milk; stir with a fork to make a soft dough. Stir only until combined.
6. Turn out dough on a lightly floured surface; knead about 10 times or until no longer sticky. Roll out dough to about 1/2 inch thick. Cut dough with a floured, round 1-1/2-inch cutter.
7. Remove casserole from oven. Increase oven heat to 400F (205C). Remove cover; arrange biscuits on top of casserole. Brush biscuits with 2 tablespoons milk; sprinkle with sesame seeds, if desired.
8. Bake, uncovered, 20 to 25 minutes or until biscuits are golden brown. Serve immediately. Makes 4 servings.

Variation
Ground pork, veal or turkey may be used in place of beef.

Top to bottom: Beef-Noodle Casserole, Biscuit-Topped Beef Bake

Lamb Stew & Herbed Dumplings

2 lbs. lamb stew cubes
Salt
Freshly ground black pepper
2 tablespoons vegetable oil
1 onion, sliced
1 garlic clove, crushed
1 red bell pepper, sliced
1 tablespoon all-purpose flour
1/2 cup beef stock
1 cup cider or apple juice
1 teaspoon Italian seasoning
1 medium eggplant, cut into 2-inch cubes

Dumplings:
1 cup cake flour
2 teaspoons baking powder
1/2 teaspoon salt
1 teaspoon Italian seasoning
1 egg
Milk

---

1. Preheat oven to 350F (175C).
2. Season lamb cubes with salt and pepper. Heat oil in a large skillet over medium heat. Add lamb cubes; sauté until browned. With a slotted spoon, transfer to a 2-1/2-quart casserole.
3. Add onion, garlic and red pepper to fat remaining in skillet; sauté until onion is transparent.
4. Stir in flour; cook 1 minute, stirring constantly. Gradually stir in stock and cider or apple juice. Boil 2 minutes, stirring constantly. Remove from heat.
5. Season with salt and pepper. Add Italian seasoning and eggplant. Spoon eggplant mixture over browned lamb cubes. Cover casserole with foil or lid. Bake in preheated oven 1-1/4 hours or until lamb is almost tender.
6. To make dumplings, combine cake flour, baking powder, salt and Italian seasoning. Beat egg in measuring cup; add enough milk to make 1/2 cup. Stir egg mixture into dry ingredients. Add additional milk, if necessary, but keep batter stiff.
7. Skim excess fat from surface of casserole. Dip a tablespoon into hot water; drop dumplings by heaping tablespoonfuls on hot lamb mixture. Cover casserole; bake 10 to 15 minutes or until dumplings are cooked. Makes 4 to 6 servings.

Lamb Chops with Apricots

4 lamb-loin double chops
Salt
Freshly ground pepper
1 tablespoon vegetable oil
1 onion, chopped
Juice of 1 orange
1/2 cup beef stock
1 tablespoon paprika
2/3 cup dried apricots (about 3 to 4 oz.)

To garnish:
Mint or parsley sprigs

If loin double chops are not available, use large loin chops.

1. Preheat oven to 350F (175C).
2. Trim chops; sprinkle with salt and pepper. Heat oil in a large skillet over medium heat. Add chops; sauté until evenly browned. Transfer to a shallow casserole.
3. Add onion to fat remaining in skillet; sauté until soft. Add orange juice, stock and paprika. Season with salt and pepper. Simmer 2 minutes.
4. Arrange apricots around browned lamb chops. Pour hot juice mixture over apricots and chops. Cover casserole with foil or lid. Bake in preheated oven 40 minutes or until lamb is tender.
5. To serve, garnish with mint or parsley. Makes 4 servings.

Variations
Canned apricots can be used instead of dried ones. Allow 1 cup canned apricot halves (about 6 to 8 halves with their juice). Replace 1/4 cup of stock with apricot juice.
Lamb Chops with Prunes: Use 2/3 cups uncooked prunes; bake as above. Serve with Duchess Potatoes. To make *Duchess Potatoes*, beat mashed potatoes with beaten egg until smooth. Fill a piping bag fitted with a star nozzle. Pipe whirls of potato on greased baking sheets. To bake potatoes, place baking sheets on oven rack above casserole. Bake 30 minutes or until potatoes are lightly browned.

Top to bottom: Lamb Stew & Herbed Dumplings, Lamb Chops with Apricots

Middle Eastern Lamb with Rice

2 oranges
1-1/2 lbs. lamb stew cubes
Salt
Freshly ground pepper
3 tablespoons vegetable oil
1 large onion, sliced
1 garlic clove, crushed
1-3/4 cups beef stock
1/2 cup raisins
1/4 teaspoon ground coriander
1 cup uncooked long-grain white rice

To garnish:
1/3 cup toasted slivered almonds (about 1-1/2 oz.)
Orange twists
Watercress

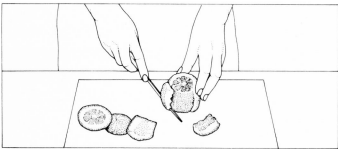
1/Cut away peel and bitter white pith.

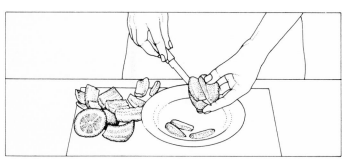
2/Cut between membrane to remove sections.

1. Grease a 2-quart casserole; set aside. Preheat oven to 350F (175C).
2. Remove peel and pith from 1 orange with a sharp knife. Remove pith from inside of peel; discard pith. Cut peel into julienne strips. Section orange. Squeeze juice from second orange.
3. Sprinkle lamb cubes lightly with salt and pepper.
4. Heat oil in a large skillet over medium heat. Add lamb cubes; sauté until lightly browned. With a slotted spoon, remove browned lamb from skillet; set aside.
5. Add onion to fat remaining in skillet; sauté until lightly browned. Add julienned orange peel, orange juice, stock and raisins; bring mixture to a boil. Add coriander, salt and pepper.
6. Add rice and browned lamb to skillet; stir well. Spoon mixture into greased casserole. Cover casserole with foil or lid.
7. Bake in preheated oven 1 hour or until rice and lamb are tender.
8. To serve, add orange sections; stir lightly. Sprinkle top of casserole with slivered almonds; garnish with orange twists and watercress. Makes 4 servings.

Variation
Substitute beef-sirloin cubes for lamb.

Lamb & Tomato Casserole

2 tablespoons vegetable oil
6 lamb shoulder chops, trimmed
1 cup chopped onions
1 garlic clove, crushed
1 cup beef stock
1/4 cup pale dry sherry
1 teaspoon Worcestershire sauce
1/2 teaspoon dried leaf basil
2 cups chopped fresh tomatoes
Salt
Freshly ground pepper

To garnish:
Green-pepper rings

1. Preheat oven to 350F (175C).
2. Heat oil in large skillet over medium heat. Add lamb chops; sauté until browned. Using tongs, place chops in a single layer in a 2-quart shallow casserole.
3. Add onions and garlic to fat remaining in skillet; sauté until transparent. Add beef stock and sherry; bring mixture to a boil. Stir in Worcestershire sauce, basil and tomatoes. Season with salt and pepper. Spoon mixture over chops.
4. Cover casserole with foil or lid; bake in preheated oven 1 hour or until lamb is cooked. Spoon any fat from surface of casserole. To serve, garnish with green-pepper rings. Makes 4 to 6 servings.

Left to right: Middle Eastern Lamb with Rice, Lamb & Potato Casserole

Lamb & Potato Casserole

2 tablespoons vegetable oil
1-1/2 lbs. lamb stew cubes
2 onions, sliced
2 carrots, sliced
1 tablespoon all-purpose flour
1/2 cup white wine
1 cup beef stock
1 tablespoon tomato paste
1 tablespoon brown sugar
Salt
Freshly ground pepper
1 to 1-1/2 teaspoons dried dill weed
1-1/2 lbs. potatoes, peeled, thinly sliced (about 4 cups)
1 tablespoon butter or margarine, melted

Dill has a strong distinctive flavor; if you prefer a less-definite taste, use less dill.

1. Preheat oven to 350F (175C).
2. Heat oil in a large skillet over medium heat. Add lamb cubes; sauté until browned. With a slotted spoon, transfer lamb cubes to a large casserole.
3. Add onions and carrots to fat remaining in skillet; sauté 2 minutes. Stir in flour; cook 1 minute, stirring frequently. Gradually stir in wine and stock. Bring mixture to a boil, stirring frequently. Remove skillet from heat.
4. Stir in tomato paste and sugar; season with salt and pepper. Spoon onion and carrot mixture over browned lamb. Stir in dill weed.
5. Arrange sliced potatoes evenly over contents of casserole. Brush potatoes with melted butter or margarine. Cover with foil or lid.
6. Bake in preheated oven 1-1/2 hours. Increase oven heat to 400F (205C). Remove cover from casserole; bake 30 minutes longer or until potatoes are tender and golden brown. Makes 4 servings.

Leg of Lamb with Garlic & Rosemary

1 (4-lb.) leg of lamb
3 to 4 garlic cloves, slivered
Few rosemary sprigs or 1 tablespoon dried
  leaf rosemary
Salt
Freshly ground pepper
3/4 lb. small white onions
2 lbs. potatoes, peeled, diced (about 6 cups)
Water
1 cup beef stock, if desired

To garnish:
Fresh rosemary sprigs

1. Preheat oven to 350F (175C).
2. Make deep cuts over surface of lamb with a small pointed knife.
3. Stick garlic slivers and small pieces of rosemary into cuts. If dried rosemary is used, sprinkle it over lamb after spiking with garlic.
4. Place lamb in a roasting pan. Roast in preheated oven 1 hour.
5. In 2 medium saucepans, cook potatoes and onions separately in boiling water 10 to 15 minutes or until almost tender. Drain well; set aside.
6. Remove lamb from oven. Arrange precooked potatoes and onions around lamb; turn in pan drippings to coat. Sprinkle potatoes and onions with salt and pepper.
7. Roast lamb 45 to 60 minutes longer or until lamb reaches desired degree of doneness. Serve garnished with fresh rosemary. A sauce can be made by deglazing roasting pan with beef stock, if desired. Degrease sauce; serve in a gravy boat or small pitcher. Makes 6 to 8 servings.

Clockwise from bottom: Crispy Pork-Loin Roast, Leg of Lamb with Garlic & Rosemary, Lamb Chops with Fennel

Lamb Chops with Fennel

4 thick lamb-shoulder arm chops
Salt
Freshly ground pepper
2 teaspoons ground coriander
1 tablespoon vegetable oil
1 onion, thinly sliced
2 fennel bulbs, chopped
1 cup beef stock
2 teaspoons cornstarch
Water

Topping:
3/4 cup fresh breadcrumbs
1/4 cup grated Cheddar cheese (1 oz.)
1 tablespoon grated Parmesan cheese

To garnish:
Fresh fennel or parsley sprigs

If chops are small or thin, allow 2 per serving; decrease cooking time by 5 to 10 minutes.

1. Preheat oven to 350F (175C).
2. Sprinkle lamb chops with salt, pepper and coriander. Heat oil in a large skillet over medium heat. Add chops; sauté until browned. With tongs, remove chops from skillet; set aside.
3. Add onion to fat remaining in skillet; sauté until soft. Drain any excess fat from skillet.
4. Place onion and fennel in a 9-inch-square baking pan; arrange browned chops over vegetables.
5. Pour stock into same skillet; bring to a boil. Pour hot stock over chops and vegetables. Cover pan with foil or lid.
6. Bake in preheated oven 45 to 55 minutes or until lamb is tender. Fennel should not be too soft.
7. Strain cooking juices into a medium saucepan; remove any fat from surface. In a small bowl, combine cornstarch and a little cold water. Stir cornstarch mixture into juices from pan; bring to a boil. Cook until thickened, stirring constantly. Pour thickened juices over vegetables and chops. Preheat broiler.
8. Combine breadcrumbs and cheeses; sprinkle cheese mixture over chops. Place lamb under preheated broiler; broil until topping is golden brown and crisp. Garnish with sprigs of fennel or parsley.

Variation
Pork chops or pork shoulder steaks may be used in place of lamb chops. Fennel has a distinctive taste. If desired, it can be replaced with a mixture of sliced onions, potatoes and carrots. Use 2 onions, 1 pound potatoes and 2 carrots.

Crispy Pork-Loin Roast

1 (2-1/2- to 3-lb.) pork-loin roast, chined or chopped
Salt
Freshly ground pepper

Topping:
1 cup fresh breadcrumbs
2 teaspoons chopped fresh sage or 1/2 teaspoon rubbed sage

To garnish:
Fresh mint or parsley sprigs

---

1. Preheat oven to 350F (175C).
2. Trim roast; sprinkle lightly with salt and pepper. Place in a roasting pan.
3. Roast in preheated oven about 1 hour.
4. Remove roast from oven; baste with pan drippings. In a small bowl, combine breadcrumbs and sage. Season crumb mixture with salt and pepper. Press crumb mixture over top of roast.
5. Increase oven heat to 400F (205C). Roast pork 30 minutes longer or until topping is crisp and lightly browned. Pork should reach an internal temperature of 170F (75C).
6. Serve roast sliced, or as a whole roast, surrounded by vegetables. Garnish with fresh mint or parsley. Makes 4 to 6 servings.

Lasagna

1 lb. green lasagna noodles
Salt
Water
1 lb. ground beef or lamb
1 large onion, chopped
2 carrots, chopped
1 (16-oz.) can tomatoes
1/2 cup beef stock
1 tablespoon tomato paste
Freshly ground pepper
2 to 3 tablespoons chopped fresh basil or
   1-1/2 teaspoons dried leaf basil
6 tablespoons dairy sour cream
1 cup grated Cheddar cheese (about 4 oz.)
1 tablespoon grated Parmesan cheese

To garnish:
Tomato slices
Cucumber slices
Parsley sprigs

1. Cook lasagna noodles in boiling salted water according to package directions until just tender. Drain on paper towels. Preheat oven to 375F (190C). Grease a 9-inch-square baking pan; set aside.
2. With no extra fat added, sauté beef or lamb in a large, heavy saucepan over medium heat until no longer pink. Add onion and carrots; cook 2 minutes. Drain excess fat from saucepan.
3. Add tomatoes, stock, tomato paste and basil; bring to a boil. Simmer meat sauce 10 minutes or until thickened. Season with salt and pepper.
4. Place a layer of lasagna noodles in greased baking pan. Cover with half the meat sauce. Repeat layers of noodles and sauce, ending with noodles.
5. Spread sour cream over noodles. Combine cheeses; sprinkle over sour cream.
6. Bake in preheated oven 45 minutes or until top is golden and sauce is bubbling. Garnish with tomato, cucumber and parsley. Makes 6 servings.

Variation
Replace basil with oregano or marjoram; add 4 ounces chopped mushrooms with tomatoes.

Left to right: Lasagna, Lamb Marsala

Lamb Marsala

1 (3- to 4-lb.) lamb-shoulder or lamb-leg roast, boneless
Salt
Freshly ground pepper

Stuffing:
2 tablespoons butter or margarine
1 onion, finely chopped
1 celery stalk, finely chopped
1-1/2 cups fresh breadcrumbs
1/3 cup chopped toasted almonds (1-1/2 oz.)
Grated peel of 1/2 lemon
1 egg, beaten

Marsala sauce:
3-1/2 tablespoons all-purpose flour
1 cup beef stock
1/4 cup Marsala
2 tablespoons lemon juice

To garnish:
Celery leaves
1/4 cup toasted flaked almonds

1. Preheat oven to 325F (165C).
2. Unroll lamb roast if it is rolled and tied; flatten if necessary. Sprinkle lightly with salt and pepper.
3. To make stuffing, melt butter or margarine in a large skillet over medium heat. Add onion and celery; sauté until soft. In a medium bowl, combine sautéed onion and celery with breadcrumbs, chopped almonds, lemon peel and egg. Season with salt and pepper.
4. Spread stuffing over flattened roast; roll lightly to enclose stuffing. Tie roast with string; place in a roasting pan.
5. Roast in preheated oven 1-1/2 to 2 hours or to desired doneness (35 to 40 minutes per pound for medium). Remove roast from pan; cover with foil to keep warm.
6. To make sauce, stir flour into pan juices; cook 1 minute, stirring constantly. Gradually stir in stock and Marsala; simmer 3 to 4 minutes.
7. Add lemon juice to hot Marsala sauce. Season with salt and pepper. Pour sauce into a gravy boat or small pitcher.
8. Remove string from roast. Cut roast into thick slices; arrange on a hot platter. Garnish with celery leaves; sprinkle with flaked almonds. Serve sauce separately. Makes 6 to 8 servings.

Fruited Pork Tenderloin

2 pork-loin tenderloins, about 1-1/2 to 2 lbs. total
Salt
Freshly ground pepper

Stuffing:
1/3 cup pitted prunes, chopped
1 cup fresh breadcrumbs
1 small onion, finely chopped
1/2 teaspoon dried leaf thyme
1 egg yolk

Sauce:
2 tablespoons vegetable oil
1 onion, chopped
1/2 cup beef stock
1/2 cup white wine
2/3 cup pitted prunes
2 tablespoons brandy, if desired
3 to 4 tablespoons whipping cream, if desired

To garnish:
Fresh mint or parsley sprigs

1. Preheat oven to 350F (175C).
2. Split pork tenderloins halfway through lengthwise with a sharp knife; flatten slightly. Season tenderloins with salt and pepper.
3. To make stuffing, combine prunes, breadcrumbs, onion and thyme. Season mixture with salt and pepper; stir in egg yolk. Spread stuffing over 1 tenderloin; cover with second tenderloin. Tie into a roast with string.
4. Heat oil in a large skillet over medium heat; sauté roast until browned. Transfer to a shallow baking pan.
5. Add onion to fat remaining in skillet; sauté until lightly browned. Add stock, wine and prunes to skillet. Season with salt and pepper; bring mixture to a boil.
6. If desired, in a small saucepan, warm brandy. Pour warmed brandy over browned roast; ignite brandy carefully. When flames die, pour sauce over and around roast; cover pan with foil or lid.
7. Roast in preheated oven 45 to 60 minutes or until roast reaches an internal temperature of 170F (75C).
8. Remove string from roast; cut into slices. Place slices in a serving dish. Stir cream into cooking sauce, if desired. To serve, spoon prunes and cooking sauce over pork slices. Garnish with mint or parsley. Makes 4 servings.

Pork Tenderloin Calvados

1-1/2 lbs. pork-loin tenderloin, cut into 3/4-inch slices
Salt
Freshly ground pepper
2 tablespoons butter or margarine

Sauce:
1 tablespoon butter or margarine
6 oz. mushrooms, sliced
3 tablespoons Calvados or other apple brandy
1/2 cup chicken stock
1/2 cup whipping cream
2 teaspoons cornstarch

To garnish:
1/3 cup toasted coarsely chopped hazelnuts
1 to 2 teaspoons chopped fresh thyme
1 tablespoon chopped fresh parsley

1. Preheat oven to 350F (175C).
2. Sprinkle pork slices lightly with salt and pepper. Melt 2 tablespoons butter or margarine in a large skillet over medium heat. Add pork slices; sauté until lightly browned. Transfer to a 2-quart casserole.
3. For sauce, add 1 tablespoon butter or margarine to skillet. Add mushrooms; sauté gently 2 to 3 minutes.
4. Add brandy, stock, salt and pepper; boil mixture 2 minutes.
5. Pour sauce mixture over pork; cover casserole with foil or lid. Bake in preheated oven 30 minutes or until pork is tender.
6. In a small bowl, combine cream and cornstarch; stir cornstarch mixture into hot casserole. Bake 5 to 10 minutes longer or until sauce is thickened.
7. Combine hazelnuts with thyme and parsley; sprinkle over casserole. Makes 4 servings.

Variation
Pork-loin tenderloin is now easily obtainable from larger supermarkets. It is a fairly expensive cut, but is solid meat, with no fat and no waste. However, if you cannot find pork-loin tenderloin, substitute boneless chicken or turkey breasts. Chicken breasts can be split in half if they are very thick, or cooked whole. Cut turkey breasts into slanting slices about 3/4 inch thick. Cook turkey or chicken in the same way as for pork.

Calvados is not always readily available. Brandy can be used instead or try using 2 tablespoons of any orange liqueur and 1 tablespoon lemon juice; this combination will alter the flavor but makes a tasty alternative.

Left to right: Fruited Pork Tenderloin, Pork Tenderloin Calvados

Curried Veal with Apple Slices

2 tablespoons all-purpose flour
2 to 3 teaspoons curry powder
1-1/2 lbs. veal stew cubes
1/4 cup vegetable oil
1 cup chopped onions
1 green apple, peeled, cored, sliced
1 garlic clove, crushed
1 tablespoon brown sugar
2 tablespoons raisins
1-1/2 tablespoons Worcestershire sauce
1 tablespoon shredded coconut
1 cup water
Salt
Freshly ground pepper

1. In a plastic bag, combine flour and curry powder to taste. Add veal cubes; shake to coat.
2. Heat oil in large skillet over medium heat. Add seasoned veal cubes; sauté until lightly browned. Remove browned veal with slotted spoon; set aside.
3. Add onions, apple and garlic to fat remaining in skillet; sauté until slightly softened. Stir brown sugar, raisins, Worcestershire sauce, coconut and water into mixture in skillet; bring to a simmer. Return browned veal to skillet; cover skillet.
4. Simmer veal mixture 45 minutes or until veal is tender, stirring occasionally. Season with salt and pepper. Makes 4 servings.

Veal in Madeira Sauce

2 tablespoons butter or margarine
1 tablespoon vegetable oil
1-1/2 lbs. veal stew cubes
8 small white onions
2 large carrots, diced
1 small red bell pepper, sliced
2 tablespoons all-purpose flour
1-1/2 cups chicken stock
1/4 cup Madeira
Salt
Freshly ground black pepper
4 oz. mushrooms, sliced
2 bay leaves

1. Preheat oven to 350F (175C).
2. Heat butter or margarine and oil in a large skillet over

medium heat. Add veal cubes; sauté until lightly browned. With a slotted spoon, transfer browned veal cubes to a 2-1/2-quart casserole.
3. Add onions and carrots to fat remaining in skillet; sauté until vegetables begin to brown. Add red pepper; cook about 2 minutes.
4. Stir in flour; cook 1 minute. Gradually stir in stock and Madeira; bring to a boil. Season sauce with salt and pepper; add mushrooms and bay leaves. Pour hot sauce over browned veal.
5. Cover casserole tightly with foil or lid; bake in preheated oven 1-1/4 to 1-1/2 hours or until veal is tender.
6. Discard bay leaves; serve with hot cooked rice. Makes 4 servings.

Veal in Wine & Cream Sauce

6 tablespoons butter or margarine
1-1/2 lbs. veal stew cubes
2 small onions
1 bouquet garni
1 cup white wine
1 cup chicken stock
Salt
Freshly ground pepper
1/4 cup all-purpose flour
1/4 cup whipping cream
2 tablespoons brandy

To garnish:
Sautéed button mushrooms
Sautéed button onions
Watercress

1. Preheat oven to 325F (165C).
2. Melt 3 tablespoons butter or margarine in a large skillet over medium heat. Add veal cubes; sauté until browned. Transfer browned veal to a 2-quart casserole; add onions and bouquet garni to casserole. Season with salt and pepper.
3. Add wine to skillet; bring to a boil. Pour hot wine over veal.
4. Cover casserole with foil or lid; bake in preheated oven 1 hour or until veal is tender.
5. Strain liquid from casserole into a medium saucepan; discard onions and bouquet garni. Boil liquid rapidly over high heat until reduced to 1-1/2 cups. Keep veal warm.
6. Cream remaining butter or margarine with flour; gradually whisk into hot sauce; cook until thickened, stirring constantly. Simmer 3 minutes. Season sauce with salt and pepper; add cream. Reheat sauce; do not boil.
7. In a small saucepan, warm brandy. Pour warmed brandy over veal; ignite brandy carefully. When flames die, pour sauce over veal; garnish with mushrooms, onions and watercress. Makes 4 servings.

Left to right: Veal in Wine & Cream Sauce, Veal Paprika

Veal Paprika

2 tablespoons butter or margarine
2 tablespoons vegetable oil
1-1/2 lbs. veal stew cubes
2 large onions, sliced
1 tablespoon paprika
1/4 cup all-purpose flour
1 cup chicken stock
1 tablespoon tomato paste
1 tablespoon lemon juice
1 (16-oz.) can tomatoes
1/3 cup raisins
Salt
Freshly ground pepper

To garnish:
1 (2-inch) cucumber piece
4 to 6 tablespoons dairy sour cream

1. Preheat oven to 350F (175C).
2. Heat butter or margarine with 1 tablespoon oil in a large skillet over medium heat. Add veal cubes; sauté until browned. Transfer to a 2-quart casserole with a slotted spoon.
3. Add remainder of oil to skillet. Add onions; sauté until golden brown. Stir in paprika and flour; cook 1 minute, stirring constantly. Gradually stir in stock; bring to a boil.
4. Add tomato paste, lemon juice, tomatoes with their liquid and raisins; simmer 1 minute. Season sauce with salt and pepper; pour over browned veal.
5. Cover casserole tightly with foil or lid; bake in preheated oven 1-1/4 to 1-1/2 hours or until veal is tender.
6. Without peeling, coarsely grate or finely dice cucumber. Spoon sour cream over hot casserole; sprinkle with grated or diced cucumber. As an alternative topping, mix cucumber and sour cream; spoon over hot casserole. Makes 4 servings.

Apple-Stuffed Veal Roast

1 (4-lb.) veal-shoulder roast, boneless
Salt
Freshly ground pepper

Stuffing:
About 3 tablespoons butter or margarine
2 celery stalks, finely chopped
1 small onion, chopped
2 oz. mushrooms, chopped
1 small apple, peeled, cored, finely chopped
1 cup fresh breadcrumbs
1 tablespoon chopped fresh parsley
1 teaspoon dried leaf thyme
Grated peel of 1 lemon
1 egg yolk

Sauce:
1 cup white wine or chicken stock
Juice of 1 small lemon
1 tablespoon cornstarch
Water

To garnish:
6 oz. button mushrooms, sautéed
Parsley sprigs
Lemon wedges

1. Preheat oven to 350F (175C).
2. Unroll roast; flatten if necessary. Sprinkle with salt and pepper.
3. To make stuffing, melt 2 tablespoons butter or margarine in a large skillet over medium heat. Add celery and onion; sauté until soft. Add mushrooms; cook 2 to 3 minutes.
4. Remove skillet from heat; add apple, breadcrumbs, parsley, thyme and half the lemon peel. Season mixture with salt and pepper; stir in egg yolk.
5. Spread stuffing over flattened roast; roll roast carefully to enclose stuffing. Tie roast with string.
6. Rub roast with butter or margarine. Place in a roasting pan.
7. Roast in preheated oven 1 hour. Baste with pan drippings; roast 1 to 1-1/2 hours longer or until roast is tender. Transfer roast to a warm serving dish.
8. Deglaze roasting pan with wine or stock. Add lemon juice and remaining lemon peel; season with salt and pepper.
9. In a small bowl, combine cornstarch with a little cold water; stir into pan juices. Boil sauce 2 minutes; season with salt and pepper. Pour sauce into a gravy boat or small pitcher.
10. To serve, slice roast; garnish with mushrooms, parsley and lemon wedges. Serve sauce separately. Makes 6 to 8 servings.

Clockwise from left: Apple-Stuffed Veal Roast, Party Ham, Veal & Ham Rolls Florentine

Party Ham

1 cook-before-eating ham half (about 4 lbs.)
1 (16-oz.) can peach halves
2 tablespoons wine vinegar
1 tablespoon lemon juice
Whole cloves
1/3 cup packed brown sugar
1 tablespoon dry mustard
1/4 teaspoon grated nutmeg
2 to 3 teaspoons cornstarch, if desired
Water, if desired

To garnish:
Watercress

1. Preheat oven to 325F (165C).
2. Place ham in a roasting pan. Drain peaches; pour juice around ham. Add vinegar, lemon juice and about 8 cloves.
3. Cover roasting pan tightly with foil or lid. Bake in preheated oven 1-1/2 hours.
4. Remove ham from oven. Increase oven heat to 400F (205C). Carefully remove any skin from ham; score fat in a diamond pattern. Baste ham with pan juices. In a small bowl, combine sugar, mustard and nutmeg; spread sugar mixture over scored fat. If desired, stud ham with cloves. Bake 30 minutes longer. Ham should reach an internal temperature of 160F (70C).
5. Stud each peach half with 3 cloves; arrange around ham. Bake an additional 10 minutes.
6. Serve surrounded with peach halves. Garnish with watercress. Pan juices can be thickened with a mixture of 2 to 3 teaspoons cornstarch blended with a little cold water, if desired. Makes 6 to 8 servings.

Hearty Oxtail Stew

1 large oxtail, cut up
1/2 cup all-purpose flour
Salt
Freshly ground pepper
2 tablespoons vegetable oil
2 large onions, sliced
2 celery stalks, sliced
2 large carrots, sliced
1 cup dark beer or ale
1 cup beef stock
1 tablespoon tomato paste
1 tablespoon brown sugar
1 tablespoon vinegar
2 bay leaves
1/4 teaspoon grated nutmeg

1. Preheat oven to 325F (165C).
2. Trim any excess fat from oxtail. Put flour in plastic bag; season with salt and pepper. Add oxtail pieces; shake to coat.
3. Heat oil in a large skillet over medium heat. Add oxtail pieces; sauté until browned. Transfer browned oxtail pieces to a large casserole or Dutch oven. Arrange vegetables around browned oxtail pieces.
4. Combine beer or ale, stock, tomato paste, sugar, vinegar, bay leaves and nutmeg in a medium saucepan. Bring mixture to a boil; season with salt and pepper. Pour over oxtail.
5. Cover casserole tightly with foil or lid; bake in preheated oven 3 hours or until oxtail is tender. Cool casserole slightly; place in refrigerator.
6. Next day, preheat oven to 325F (165C). Remove fat from top of casserole; discard bay leaves. Bake in preheated oven 1 hour before serving. Makes 4 servings.

Veal & Ham Rolls Florentine

4 (4-oz.) veal cutlets
4 slices cooked ham
Salt
Freshly ground pepper
Grated nutmeg
1 (10-oz.) pkg. frozen leaf spinach, thawed
2 tablespoons butter or margarine
2 tablespoons vegetable oil
1 onion, sliced
1/3 cup all-purpose flour
1 cup chicken stock
1/2 cup dry white wine
1/4 cup half and half
4 oz. mushrooms, quartered, sliced

To garnish:
Parsley sprigs

1. Preheat oven to 350F (175C).
2. Lay cutlets on a flat surface. Cover each with a slice of ham; sprinkle with salt, pepper and nutmeg.
3. Squeeze excess water from spinach; divide into 4 equal portions. Place 1 portion of spinach on each cutlet. Roll each cutlet to enclose spinach; secure with wooden picks.
4. Heat oil in a large skillet over medium heat. Add veal rolls; sauté until lightly browned. Place in a baking dish large enough to hold rolls in 1 layer.
5. Add onion to fat remaining in skillet; sauté until soft. Stir in flour; cook 1 minute, stirring constantly. Gradually stir in stock and wine; bring to a boil.
6. Remove sauce from heat; add half and half and mushrooms. Pour hot sauce over veal rolls. Cover baking dish with foil or lid; bake in preheated oven 1 hour.
7. Remove wooden picks; garnish with parsley. Makes 4 servings.

> If the veal cutlets are over 1/4 inch thick or are cut unevenly, pound between 2 sheets of plastic wrap or waxed paper. Use a regular meat mallet if you have one; if not, the side of a sturdy saucer or the side of an empty champagne bottle make excellent substitutes.

Medallions of Pork with Sesame Rounds

2 lbs. pork-loin tenderloin, cut into 3/4-inch slices
Salt
Freshly ground pepper
1/4 cup butter or margarine
1 tablespoon finely chopped onion
2 tablespoons all-purpose flour
1 cup milk
1/2 cup chicken stock
2 teaspoons lemon juice
2 teaspoons dried leaf tarragon or 4 teaspoons chopped
   fresh tarragon

Sesame Rounds:
1 cup all-purpose flour
1/4 teaspoon salt
2 tablespoons butter or margarine
2 tablespoons shortening
1 to 2 tablespoons ice water
1 egg yolk blended with 1 tablespoon water for glaze
Sesame seeds
Fresh tarragon or parsley

1. Preheat oven to 350F (175C).
2. Sprinkle slices of pork with salt and pepper.
3. Melt butter or margarine in a large skillet over medium heat. Add pork slices; sauté until lightly browned. With tongs, transfer browned pork to a shallow casserole.
4. Add onion to fat remaining in skillet; sauté until soft. Stir in flour; cook 1 minute, stirring constantly.
5. Gradually stir in milk and stock; bring mixture to a boil. Add lemon juice and tarragon. Season sauce with salt and pepper; pour over pork.
6. Cover casserole with foil or lid; bake in preheated oven 50 minutes or until pork is tender.
7. To make Sesame Rounds, preheat oven to 400F (205C). Grease a baking sheet; set aside.
8. In a medium bowl, combine flour and salt. Cut in butter or margarine and shortening until mixture resembles coarse crumbs. Stir ice water into flour mixture, 1 tablespoon at a time, until mixture begins to bind together. Gather dough into a ball; shape into flattened round.
9. Roll out dough on a lightly floured surface to 1/4-inch thick. Cut dough with floured 2- to 2-1/2-inch fluted round cutter. Place rounds on greased baking sheet about 1 inch apart. Prick rounds with a fork. Brush with glaze; sprinkle with sesame seeds.
10. Bake in preheated oven 10 to 12 minutes or until golden brown. Cool on a wire rack; reheat when required.
11. To serve, spoon pork and sauce into a serving dish; garnish with Sesame Rounds and fresh tarragon or parsley. Serve with Duchess Potatoes, page 19. Makes 4 servings.

Variation
Sliced turkey or chicken breasts can also be used for this recipe. If using poultry, cooking time can be decreased by 15 minutes. Try using fresh or dried rubbed sage instead of tarragon.

Medallions of Pork with Sesame Rounds

Pork Chop Casserole

6 (1/4-inch-thick) center-cut pork chops
2 tablespoons vegetable oil
2 onions, cut into wedges
1 garlic clove, crushed
1 cup beer
1 cup beef stock
3 carrots, sliced
1 tablespoon Dijon-style mustard
2 teaspoons brown sugar
Salt
Freshly ground pepper
3 white-bread slices, crusts removed
1 egg, beaten
2 tablespoons milk

To garnish:
Chopped fresh parsley

1. Preheat oven to 350F (175C).
2. Heat oil in a medium skillet over medium heat. Add pork chops; sauté until browned. With tongs, transfer browned chops to a 2-quart casserole.
3. Add onions and garlic to oil remaining in skillet; sauté until onion is transparent. Gradually add beer and beef stock; bring mixture to a boil. Stir in carrots, mustard, sugar, salt and pepper. Pour mixture over pork chops.
4. Cover casserole with foil or lid; bake in preheated oven 1 hour. Cut bread into squares or triangles. In a flat bowl, beat together egg, milk, salt and pepper; dip bread triangles or squares into mixture.
5. Remove casserole from oven; stir lightly. Preheat broiler. Place egg-soaked bread on top of casserole. Broil until bread is crisp and lightly browned. Garnish with chopped parsley. Makes 4 to 6 servings.

Clockwise from top left: Pork Chop Casserole, Pork & Potato Bake, Oriental Pork Stew

Oriental Pork Stew

1/4 cup butter or margarine
1-1/2 lbs. lean pork cubes
1 onion, finely chopped
4 celery stalks, thinly sliced
1 large red bell pepper, thinly sliced
1 to 2 garlic cloves, crushed
1 (8-oz.) can water chestnuts, drained, sliced if large
1/4 cup medium-dry sherry
2 tablespoons soy sauce
1 tablespoon lemon juice
About 3/4 cup beef stock
Salt
Freshly ground black pepper

Topping:
3 tablespoons butter or margarine
3/4 cup fresh breadcrumbs
1/3 cup chopped blanched almonds

To garnish:
Chopped fresh parsley

Pork & Potato Bake

1 tablespoon vegetable oil
1-1/2 lbs. lean pork, cut into 1/2-inch strips
2 lbs. potatoes, thinly sliced (about 6 cups)
2 medium onions, thinly sliced
6 small sweet pickles, sliced
Salt
Freshly ground pepper
6 tablespoons dry white wine
1/2 cup half and half

To garnish:
Chopped fresh parsley

1. Preheat oven to 350F (175C). Grease a 2-1/2-quart casserole; set aside.
2. Heat oil in a large skillet over medium heat. Add pork strips; sauté until browned.
3. Layer potatoes in greased casserole with browned pork, onions and pickles. Season with salt and pepper.
4. Combine wine and half and half; pour over potatoes.
5. Cover casserole with foil or lid; bake in preheated oven 1 hour. Remove casserole cover; increase oven heat to 400F (205C).
6. Bake 30 to 40 minutes longer or until potatoes are golden brown on top. Garnish with parsley. Makes 4 servings.

Variation
Any cut of lean pork can be used in this recipe, or substitute chicken or turkey breasts cut into 1/2-inch strips.

1. Preheat oven to 350F (175C).
2. Melt 2 tablespoons butter or margarine in a large skillet over medium heat. Add pork; sauté until browned. With a slotted spoon, transfer browned pork to a 2-quart casserole.
3. Add remaining butter or margarine to skillet. Add onion, celery, red pepper and garlic; cook over high heat 2 minutes, stirring frequently. Transfer vegetables to casserole containing browned pork. Stir in water chestnuts.
4. In a measuring cup, combine sherry, soy sauce and lemon juice; add enough stock to make 1 cup. Stir stock mixture into casserole; season with salt and pepper.
5. Cover casserole with foil or lid.
6. Bake in preheated oven 50 to 60 minutes or until pork is tender.
7. Meanwhile, melt butter or margarine for topping in a small skillet. Add breadcrumbs; sauté until lightly browned. Stir in nuts; cook until breadcrumbs are golden brown.
8. To serve, spoon nut topping over casserole; garnish with parsley. Makes 4 servings.

Pork Tenderloin Véronique

2 pork-loin tenderloins, trimmed
24 green grapes, halved, seeded, if necessary
1/4 cup butter or margarine
Salt
Freshly ground pepper
1 onion, quartered
1 carrot, sliced
1 bay leaf
1/4 teaspoon dried leaf thyme
1/2 cup white wine
1/2 cup chicken stock
2 tablespoons butter or margarine
2 tablespoons all-purpose flour
1 egg yolk

---

1. Preheat oven to 350F (175C).
2. Cut tenderloins halfway through lengthwise; open out. Sandwich 24 grape halves together with 1/4 cup butter or margarine; arrange filled grapes down the center of 1 tenderloin. Place second tenderloin on top to form a roast; tie with string.
3. Place tenderloin roast in a shallow roasting pan. Arrange onion, carrot and bay leaf around roast; sprinkle with thyme. Pour wine and stock around roast. Roast in preheated oven 1-1/4 hours or until done. Pork should reach an internal temperature of 170F (80C).
4. Move roast to a warm serving platter; cover with foil to keep warm. Remove fat from cooking liquid. Strain liquid into a medium saucepan.
5. Cream flour and butter to make a beurre manié; gradually whisk flour mixture into strained cooking liquid. Cook over low heat, stirring constantly, until thickened. In a small bowl, beat egg yolk lightly; stir in some of hot sauce. Return mixture to saucepan; simmer sauce 2 to 3 minutes. Pour sauce over roast; garnish with remaining grape halves. Makes 4 servings.

Ham Bake

1/4 cup butter or margarine
1/2 cup chopped onion
8 oz. mushrooms, sliced
3 tablespoons all-purpose flour
Salt
Freshly ground pepper
1/2 pint half and half (1 cup)
2 tablespoons pale dry sherry
1-1/2 to 2 cups cubed cooked ham
1 (5-oz.) can sliced water chestnuts, drained
1/4 cup shredded Swiss cheese (1 oz.)
1/2 cup fresh breadcrumbs

---

1. Preheat oven to 400F (205C). Grease a 2-quart casserole; set aside.
2. Melt 2 tablespoons butter or margarine in a large skillet. Add onion and mushrooms; sauté until onion is transparent.
3. Stir flour, salt and pepper into onion mixture. Stir in half and half and sherry; cook until mixture is thickened and bubbly, stirring frequently. Add ham and water chestnuts.
4. Spoon ham mixture into greased casserole; top with shredded cheese. Melt remaining 2 tablespoons butter or margarine in a small skillet. Stir in breadcrumbs; sprinkle buttered crumbs over casserole. Bake, uncovered, 25 minutes. Makes 4 servings.

Many people have a preference for rare, medium or well-done meat and would not consider eating meat any other way. The use of a meat thermometer ensures that meat will be cooked perfectly. Beef or lamb can be cooked to rare (140F, 60C), medium (160F, 70C) or well-done (170F, 75C). Both beef and lamb are more juicy and flavorful if not overcooked. Remove roasts from the oven when the internal temperature is 5 to 10 degrees below desired doneness. Roast will continue to cook a few minutes after it is removed from the oven.

Unlike beef or lamb, pork should not be served rare or medium-done. Cook fresh pork to an internal temperature of 170F (75C) for flavor and juiciness. Meat should be gray in color with no pink remaining. Although the parasite, trichina, is rarely found in pork today, it is easily destroyed by cooking. Usually trichina is destroyed at 140F (60C).

Stuffed Pork Chops

Stuffing:
1 cup fresh breadcrumbs
1 teaspoon dried rubbed sage
1 small onion, finely chopped
3 tablespoons raisins
Finely grated peel of 1/2 lemon
2 tablespoons lightly beaten egg
Salt
Freshly ground pepper
4 extra-thick pork chops
1 cup apple juice or white wine

To garnish:
1 tablespoon finely chopped parsley

1. Preheat oven to 350F (175C).
2. For stuffing, combine breadcrumbs, sage, onion, raisins, lemon peel and egg. Season with salt and pepper; stir well.
3. Cut a slit in side of each pork chop to make a pocket for stuffing.
4. Stuff each pork chop; arrange stuffed chops in a shallow baking pan. Pour juice or wine around stuffed chops; bake in preheated oven 1 hour or until pork is tender. To serve, arrange on serving platter; sprinkle with parsley. Makes 4 servings.

Variation
If thick pork chops are not available, use 8 thin ones. Sandwich stuffing between 2 chops; fasten with wooden picks. If desired, the cooking liquid can be thickened with a little cornstarch blended with cold water.

Left to right: Pork Tenderloin Véronique, Stuffed Pork Chops

Poultry & Game

Chicken Marengo

1 (3-lb.) broiler-fryer chicken, cut up
Salt
Freshly ground pepper
1/4 cup butter or margarine
2 tablespoons brandy
1 onion, sliced
1 to 2 garlic cloves, crushed
2 tablespoons all-purpose flour
1 (16-oz.) can tomatoes, pureed or very finely chopped
1/2 cup dry white wine
1 tablespoon tomato paste
4 oz. button mushrooms, halved

To garnish:
2 hard-cooked eggs, quartered
Black olives
4 crayfish or 8 large whole shrimp, if desired
2 tablespoons butter or margarine, if desired

1. Preheat oven to 350F (175C).
2. Season chicken with salt and pepper.
3. Melt 2 tablespoons butter or margarine in a large skillet over medium heat. Add chicken; sauté until browned. Transfer browned chicken to a 3-quart casserole.
4. In a small saucepan, warm brandy; pour warmed brandy over chicken. Ignite carefully; let burn until flame dies.
5. Add remaining butter or margarine to skillet. Add onion and garlic; sauté until lightly colored.
6. Stir flour into onions; cook 1 minute, stirring constantly. Gradually stir in tomatoes, wine and tomato paste; bring to a boil.
7. Season with salt and pepper. Add mushrooms; simmer 2 minutes. Pour over chicken. Cover casserole with foil or lid.
8. Bake in preheated oven 45 to 50 minutes or until chicken is tender.
9. Arrange chicken and sauce in a deep serving dish; garnish with hard-cooked eggs and olives. For additional garnish, if desired, lightly sauté crayfish or shrimp in butter or margarine. Makes 4 servings.

Chicken with Artichoke Hearts

4 boneless chicken breasts
Salt
Freshly ground pepper
3 tablespoons butter or margarine
1 onion, sliced
1/2 cup dry white wine
1/2 cup chicken stock
1 (10-oz.) pkg. frozen artichoke hearts, thawed
2 tablespoons all-purpose flour
6 tablespoons half and half

To garnish:
Watercress

A cut up broiler-fryer chicken may also be used for this recipe.

1. Season chicken with salt and pepper. Heat butter or margarine in a large skillet over medium heat. Add seasoned chicken; sauté until browned. With tongs, remove chicken; set aside.
2. Add onion to fat remaining in skillet; sauté until soft but only lightly colored. Add wine and stock. Return chicken to skillet.
3. Cut each artichoke heart in half; add to skillet.
4. Cover and simmer 25 to 30 minutes or until chicken is tender. With a slotted spoon, remove chicken and artichokes to a warmed serving dish; cover with foil to keep warm.
5. In a small bowl, combine flour and half and half. Whisk flour mixture into skillet. Simmer sauce 3 to 4 minutes; pour over chicken and artichokes. Garnish with watercress. Makes 4 servings.

Variation
This makes a hearty filling for a pie. Chill cooked mixture until cool. Coarsely chop chicken; return to sauce. Use pastry for a double-crust pie. Pour filling into a pastry-lined, 9-inch, deep-dish pie plate. Cover with top crust; brush with beaten egg. Bake in preheated 400F (205C) oven 45 minutes or until crust is golden.

Top to bottom: Chicken Marengo, Chicken with Artichoke Hearts

Chicken Rossini

4 boneless chicken breasts
Salt
Freshly ground pepper
About 4 oz. firm liver pâté
3 tablespoons butter or margarine
6 oz. button mushrooms, sliced
3 tablespoons brandy
6 tablespoons chicken stock
4 slices bread
2 tablespoons butter or margarine
2 tablespoons vegetable oil
3 tablespoons half and half, if desired

To garnish:
Watercress

1. Remove skin from chicken breasts; pound to flatten. Sprinkle lightly with salt and pepper.
2. Cut pâté into 4 slices; wrap each slice with a chicken breast. Tie with string or secure with wooden picks.
3. Melt 3 tablespoons butter or margarine in a large skillet over medium heat. Add chicken; sauté until lightly browned. With tongs, set browned chicken aside.
4. Add mushrooms to fat remaining in skillet; sauté 1 minute. Add brandy and stock; bring to a boil. Season mushroom mixture with salt and pepper; return chicken to skillet.
5. Cover skillet with lid; simmer over low heat 20 to 25 minutes or until chicken is tender.
6. Cut bread into ovals the size of chicken rolls. Heat 2 tablespoon of butter or margarine and oil in a medium skillet. Sauté bread ovals in hot oil mixture until golden.
7. Place sautéed bread on a hot serving dish. Top each oval with a cooked chicken roll; remove string or wooden picks. Adjust seasoning in sauce. Add half and half, if desired; reheat, but do not boil. Spoon sauce over and around chicken. Garnish with watercress. Makes 4 servings.

Chicken Rossini

Chicken Breasts with Apples & Cream

7 tablespoons butter or margarine
6 chicken-breast cutlets
2 apples, peeled, cored, sliced
2 tablespoons all-purpose flour
Salt
Freshly ground pepper
1 cup chicken stock
1/4 cup dry white wine
1/4 cup whipping cream

1. Melt 5 tablespoons butter or margarine in a large skillet over medium heat. Add cutlets; sauté 8 to 10 minutes or until chicken is tender and lightly browned, turning several times. Arrange cooked chicken in a serving dish; cover with foil to keep warm.
2. Add apples to fat remaining in skillet; cook until barely tender. Arrange apples around chicken.
3. Melt remaining 2 tablespoons butter or margarine in small saucepan. Blend in flour, salt and pepper. Gradually stir in stock; cook over medium heat until thickened, stirring constantly. Add wine; simmer 1 minute. Stir in cream; reheat sauce, but do not boil. Pour some sauce over chicken breasts. Serve additional sauce separately. Makes 4 to 6 servings.

Baked Orange Chicken

1 cup orange juice
1 tablespoon dry vermouth
1 garlic clove, crushed
Salt
Freshly ground pepper
6 chicken-breast cutlets
1/2 cup orange marmalade

To garnish:
1 (11-oz.) can mandarin oranges, drained
Toasted slivered almonds

1. To make marinade, in a small bowl, combine orange juice, vermouth, garlic, salt and pepper. Place cutlets in a single layer in a 13" x 9" glass baking dish; pour marinade over. Cover and refrigerate several hours or overnight.
2. Preheat oven to 350F (175C).
3. Remove cover from pan; spoon some marmalade on top of each cutlet. Bake, uncovered, 30 minutes or until chicken is tender, basting occasionally with marinade.
4. Place chicken on serving platter. Strain juices into small pitcher or gravy boat; serve separately. Garnish chicken with mandarin oranges and almonds. Makes 4 to 6 servings.

Mexican Chicken

3 to 4 tablespoons olive oil or vegetable oil
1 (3-1/2- to 4-lb.) broiler-fryer chicken, cut up
2 garlic cloves, finely chopped
1 (16-oz.) can tomatoes
1 (4-oz.) can chopped green chilies, drained
1 tablespoon chili powder
1 teaspoon cumin
Hot-pepper sauce
Salt
Freshly ground black pepper

To garnish:
Chopped fresh parsley

1. Heat oil in a large skillet over medium heat. Add chicken; sauté until brown on all sides. Using tongs, remove chicken from skillet; place in single layer in a shallow baking dish. Set chicken aside.
2. Preheat oven to 350F (175C).
3. Discard all but 2 tablespoons oil from skillet; add garlic; sauté 3 minutes. Add tomatoes, chilies, chili powder, cumin and hot-pepper sauce. Season with salt and pepper. Stir well; simmer 5 minutes.
4. Spoon tomato mixture over chicken. Cover casserole with foil or lid. Bake 20 minutes in preheated oven. Uncover and bake 25 minutes longer or until chicken is tender, basting occasionally.
5. To serve, arrange chicken on a serving plate; spoon sauce over top. Sprinkle with parsley. Serve with cooked rice and refried beans. Makes 4 servings.

To make *chicken cutlets*, place boned chicken-breast halves on a work surface between sheets of waxed paper or plastic wrap. Using a meat mallet or cleaver, pound each breast until about 1/4 inch thick. To make turkey cutlets, cut boneless turkey breast into 1/2-inch diagonal slices before pounding. Chicken and turkey cutlets can be frozen on baking sheets, wrapped and stored up to 2 months.

Chicken Patties with Mustard Sauce

1 lb. boneless chicken breasts, skinned
2 cups fresh breadcrumbs
1/2 cup butter or margarine
1/2 cup whipping cream
Salt
Freshly ground pepper
1/4 teaspoon ground nutmeg

Mustard sauce:
2 tablespoons butter or margarine
1 green onion, thinly sliced
2 tablespoons all-purpose flour
2 to 3 teaspoons Dijon-style mustard
1 cup chicken stock

1. In a food processor fitted with a steel blade, process chicken until coarsely chopped. Add 1 cup breadcrumbs, 1/4 cup butter or margarine, cream, salt, pepper and nutmeg; process until chicken is finely minced, stopping machine to scrape down sides. Chill chicken mixture at least 30 minutes.
2. Shape into 6 (1/2-inch-thick) patties. Dip patties into remaining 1 cup breadcrumbs, pressing to coat well.
3. Melt remaining 1/4 cup butter or margarine in a large skillet. Sauté patties 5 minutes on each side or until lightly browned and crisp.
4. To make mustard sauce, melt butter or margarine in a medium saucepan. Add onion; sauté until tender. Stir flour and mustard into saucepan; stir until blended. Gradually add stock, stirring constantly; cook until mixture is thickened and smooth.
5. Arrange chicken patties on a platter. Top with mustard sauce. Makes 4 to 6 servings.

Beef & Game Casserole

1 to 2 Cornish hens, quartered
12 oz. beef stew cubes
Salt
Freshly ground pepper
2 tablespoons butter or margarine
1 tablespoon vegetable oil
8 oz. small white onions
1 tablespoon all-purpose flour
1-1/2 cups beef stock
1/2 cup red wine
4 oz. button mushrooms, trimmed, halved
1 bay leaf
2 to 3 tomatoes, peeled, quartered, if desired

Caraway balls:
2 cups fresh breadcrumbs
2 tablespoons vegetable oil
1/4 to 1/2 teaspoon caraway seeds
2 tablespoons lightly beaten egg
Vegetable oil for frying

1. Preheat oven to 350F (175C).
2. Season hens and beef cubes with salt and pepper.
3. Heat butter or margarine and oil in a large skillet over medium heat. Add hens; sauté until browned. Transfer to a 3-quart casserole or Dutch oven. Add beef cubes; sauté until browned. With a slotted spoon, transfer browned cubes to casserole with hens.
4. Add onions to fat remaining in skillet; sauté until soft. Stir in flour; cook 1 minute.
5. Gradually stir in stock and wine; bring to a boil. Add mushrooms, bay leaf and tomatoes, if desired. Pour hot mixture over beef and hens.
6. Cover casserole tightly with foil or lid. Bake in preheated oven 1-3/4 hours or until beef and hens are tender.
7. To make caraway balls, combine breadcrumbs, 2 tablespoons oil, caraway seeds, salt and pepper in a medium bowl. Stir in beaten egg. Shape mixture into 4 balls. Heat oil in a medium skillet; add caraway balls. Sauté 5 minutes or until golden brown.
8. Discard bay leaf from casserole; serve topped with caraway balls. Makes 4 servings.

Glazed Roast

1 (4-lb.) pork-shoulder blade or venison roast, boneless
Salt
Freshly ground pepper

Glaze:
5 to 6 tablespoons dry sherry
3 tablespoons red-currant jelly
2 teaspoons lemon juice

To garnish:
Cooked artichokes
Watercress

1. Preheat oven to 325F (165C).
2. Season roast with salt and pepper; place in a roasting pan.
3. Roast in preheated oven 2-1/2 to 3 hours or until roast reaches an internal temperature of 170F (75C).
4. To make glaze, combine sherry, jelly and lemon juice in a small saucepan; place over low heat until jelly melts. Brush over roast 3 to 4 times during baking.
5. To serve, place roast on a warmed serving plate; remove string. Let roast stand 10 to 15 minutes before carving; garnish with artichokes and watercress. Makes 6 to 8 servings.

The term *game* refers to all wild animals used for food. This includes wild birds, small game, such as squirrels and rabbits, and large game, such as deer, elk and moose. Meat from deer is called venison. Game is not widely available unless friends or family members are hunters.

In general, game tends to have less fat than domestic animals. Marinating, basting with drippings or larding with bacon or pork fat are several ways of making game more moist and juicy. *Larding* is the insertion of long, thin strips of fat into meat with a special device called a *larding needle*.

Much of the gamey flavor of venison comes from the fat. If desired, remove most of the venison fat; substitute beef fat, pork fat or bacon.

Game birds have dark meat. Older game birds need long, slow braising. They are excellent in stews and casseroles. Young birds require shorter cooking times.

Left to right: Beef & Game Casserole, Glazed Roast

Marinated Chicken

1 (2-1/2- to 3-lb.) broiler-fryer chicken, quartered

Marinade:
1 onion, sliced
2 tablespoons vegetable oil
6 peppercorns
1 tablespoon lemon juice
1/2 cup red wine

Casserole:
2 tablespoons vegetable oil
2 onions, chopped
2 celery stalks, sliced
2 carrots, sliced
1-1/2 to 2 tablespoons all-purpose flour
1 cup beef stock
1 tablespoon red-currant jelly
2 bay leaves
Salt
Freshly ground pepper

Stuffing balls:
2 tablespoons butter or margarine
1 small onion, grated
2 celery stalks, finely chopped
Grated peel of 1/2 lemon
Pinch of grated nutmeg
2 cups fresh breadcrumbs
1 tablespoon chopped fresh parsley
1 egg yolk

To garnish:
Grilled bacon rolls
Parsley

The marinade will give chicken a mild game-bird flavor that is a delightful change from the usual baked chicken.

1. Place chicken pieces in a glass or ceramic bowl; add marinade ingredients. Cover and refrigerate 3 hours.
2. Preheat oven to 325F (165C). Drain chicken; pat dry with paper towels. Strain marinade; discard onion and peppercorns. Heat oil in a large skillet over medium heat. Add chicken; sauté until brown. With tongs, transfer browned chicken to a large casserole or Dutch oven.
3. Add vegetables to fat remaining in skillet; sauté gently until golden brown, stirring occasionally. Sprinkle vegetables with flour; cook 1 minute, stirring constantly. Gradually stir in strained marinade, stock, jelly, bay leaves, salt and pepper. Pour mixture over chicken.
4. Cover casserole tightly with foil or lid. Bake in preheated oven 1-1/2 hours or until chicken is tender.
5. To make stuffing balls, grease an 8-inch-square baking pan; set aside. Heat butter or margarine in a large skillet. Add onion and celery; sauté gently until soft. Stir in remaining ingredients. Shape mixture into balls. Place in greased pan; bake 30 minutes. Bake stuffing during chicken's final 30 minutes of baking.
6. Remove bay leaves and any fat from surface of casserole. Spoon into a serving dish; top with baked stuffing balls. Garnish with bacon rolls and parsley. Makes 4 to 6 servings.

Marinated Chicken

Duck with Red Cabbage

1 (5- to 6-lb.) duck, thawed, if frozen
Salt
Freshly ground pepper
8 oz. Italian sausages, halved

Cabbage:
2 tablespoons butter or margarine
2 onions, sliced
Bouquet garni
1/2 cup red wine
1/2 cup beef stock
2 tablespoons vinegar
1 head red cabbage, finely shredded
Salt
Freshly ground pepper

To garnish:
Parsley sprigs
Lemon wedges

1. Preheat oven to 350F (175C).
2. Remove giblets and neck from duck; reserve for other use. Remove excess fat from duck and discard. Prick skin all over to allow fat to escape during roasting. Place duck, breast-side up, on a rack in a shallow roasting pan. Sprinkle duck with salt and pepper.
3. Roast duck in preheated oven 1-1/2 hours. Arrange sausages around duck. Roast 30 to 40 minutes longer or until duck is done and sausages are cooked. Juices should run clear when duck is pierced between breast and thigh.
4. Prepare cabbage while duck is roasting. Melt butter or margarine in a large saucepan over medium heat. Add onions; sauté until soft.
5. Add bouquet garni, wine, stock and vinegar to saucepan; bring to a boil. Add cabbage; season with salt and pepper. Cover and cook over low heat 20 minutes or until cabbage is tender. Discard bouquet garni.
6. Cut duck into quarters. Arrange cabbage in a deep serving dish; top with duck and sausages. Garnish with parsley and lemon. Makes 4 servings.

Duck with Plum Sauce

1 (5- to 6-lb.) duck, thawed, if frozen
Salt
Freshly ground pepper
Garlic powder
1 (1-lb.) can plums in syrup
2 teaspoons Worcestershire sauce
1 tablespoon wine vinegar

To garnish:
Watercress
Sautéed cherry tomatoes

1. Preheat oven to 350F (175C).
2. Remove giblets and neck from duck; reserve for other use. Remove excess fat from duck; discard. Prick skin all over to allow fat to escape during roasting. Place duck, breast-side up, in a shallow roasting pan. Sprinkle duck with salt, pepper and garlic powder.
3. Roast in preheated oven 1-1/2 hours.
4. Spoon off all fat from pan, leaving pan juices around duck.
5. Meanwhile, make plum sauce. Remove pits from plums. In a blender or a food processor fitted with a steel blade, puree plums with syrup until smooth. Blend in Worcestershire sauce, vinegar, salt and pepper.
6. Pour plum sauce over duck; bake 30 minutes longer or until duck is done. Juices should run clear when duck is pierced between breast and thigh.
7. Transfer duck to a serving dish; cover with foil to keep warm. Spoon fat off pan juices; adjust seasoning. Pour into a gravy boat or small pitcher. To serve, garnish duck with watercress and cherry tomatoes; accompany with sauce. Makes 4 servings.

Turkey & Bean Bake

2 tablespoons butter or margarine
1 tablespoon vegetable oil
5 cups diced uncooked turkey
1 (15-oz.) can dried lima beans, drained
1 (15-oz.) can red kidney beans, drained
1 large onion, sliced
1 garlic clove, crushed
1 red bell pepper, chopped
1 tablespoon all-purpose flour
1/2 cup white wine
1 (16-oz.) can tomatoes
1/2 cup chicken stock
1 teaspoon ground ginger
1 tablespoon soy sauce
1 tablespoon Worcestershire sauce
Salt
Freshly ground black pepper

1. Preheat oven to 350F (175C).
2. Heat butter or margarine and oil in a large skillet over medium heat. Add turkey; sauté until lightly browned. With a slotted spoon, transfer browned turkey to a 3-quart casserole.
3. Add lima and kidney beans to turkey.
4. Add onion and garlic to fat remaining in skillet; sauté until onion is soft. Add red pepper; cook 2 minutes.
5. Stir in flour; cook 1 minute, stirring constantly. Stir in wine, tomatoes and stock; bring to a boil.
6. Add tomato mixture to turkey and beans. Stir in ginger, soy sauce and Worcestershire sauce. Season with salt and pepper. Cover casserole with foil or lid; bake in preheated oven 1 hour. Makes 6 servings.

Duck with Orange Sauce

1 (5-to 6-lb.) duck, quartered
Salt
Freshly ground pepper
2 oranges
1 cup beef stock
Juice of 1/2 lemon
2 tablespoons orange marmalade
1 tablespoon cornstarch
1/4 cup port or red wine

To garnish:
Orange sections
Watercress

1. Preheat oven to 375F (190C).
2. Trim duck; remove excess fat. Prick skin all over with a fork to allow fat to escape during roasting. Sprinkle lightly with salt and pepper.
3. Place duck quarters, skin-side up, on a rack in a shallow roasting pan. Roast in preheated oven 1 hour or until duck is done. If skin is not crispy, place duck under broiler a few minutes.
4. With a vegetable peeler, remove peel from oranges. Cut peel into julienne strips. Or, coarsely grate peel from oranges. Juice both oranges. Set peel and juice aside.
5. Remove roasted duck to a warmed serving platter; cover with foil to keep warm. Remove fat from pan drippings. Deglaze roasting pan with stock. Stir orange juice and peel, lemon juice and marmalade into roasting pan.
6. Bring sauce to a boil; simmer until marmalade has melted. In a small bowl, combine cornstarch and port or wine. Stir cornstarch mixture into hot sauce. Cook until sauce is thickened, stirring constantly.
7. Spoon some sauce over duck; garnish with orange sections and watercress. Serve remaining sauce separately. Makes 4 servings.

Clockwise from top: Duck with Plum Sauce, Duck with Orange Sauce, Duck with Red Cabbage

1/Cut in half, breast-side up.

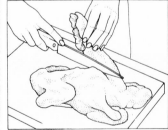

2/Remove backbone.

3/Cut each half in two.

Clockwise from top left: Italian-Style Turkey, Turkey Fricassee, Chicken Véronique

Italian-Style Turkey

4 thick turkey-breast fillets
Salt
Freshly ground pepper
Garlic powder
2 tablespoons butter or margarine
1 (16-oz.) can tomatoes
1 tablespoon tomato paste
1/4 cup vermouth
2 teaspoons lemon juice
2 bay leaves

To garnish:
Parsley sprigs
Lemon slices

1. Sprinkle turkey fillets with salt, pepper and garlic powder.
2. Melt butter or margarine in a large heavy saucepan over medium heat. Add fillets; sauté until lightly browned. Remove browned fillets; set aside.
3. In a blender or food processor fitted with a steel blade, puree canned tomatoes. Pour pureed tomatoes and tomato paste into saucepan; cook until tomatoes are reduced by almost half.
4. Add vermouth and lemon juice; bring mixture to a boil. Season with salt, pepper and garlic powder. Return fillets to saucepan; add bay leaves.
5. Cover and simmer over low heat 20 to 30 minutes or until turkey is tender.
6. To serve, discard bay leaves. Place turkey fillets on a deep serving platter; spoon sauce over fillets. Garnish with parsley and lemon slices. Serve with hot cooked rice or pasta. Makes 4 servings.

Turkey Fricassee

5 cups diced uncooked turkey breast
2 onions, chopped
3 carrots, sliced
1 bay leaf
2 cups chicken stock
Salt
Freshly ground pepper
1/4 cup butter or margarine
4 oz. button mushrooms, sliced
1/3 cup all-purpose flour
1 egg yolk
3 tablespoons whipping cream
Juice of 1/2 small lemon

To garnish:
Grilled bacon rolls, page 8
Watercress

1. Place turkey, onions, carrots, bay leaf and stock in a large saucepan over medium heat.
2. Season turkey mixture with salt and pepper. Bring to a boil. Cover and simmer 15 minutes or until turkey and vegetables are done.
3. Strain cooking liquid from turkey; reserve. Discard vegetables and bay leaf. Keep turkey warm.
4. Melt butter or margarine in a medium saucepan. Add mushrooms; sauté 1 minute. Stir in flour; cook 1 minute, stirring constantly. Gradually stir in reserved cooking liquid; bring to a boil. Cook until thickened, stirring frequently.
5. In a small bowl, blend egg yolk with cream. Stir in a little of sauce; stir egg-yolk mixture into hot sauce. Reheat sauce, but do not boil. Adjust seasoning; stir in lemon juice.
6. Pour sauce over turkey; stir well. Garnish with bacon rolls and watercress.

Variation:
This recipe can be made equally well with veal or pork. Use veal cubes or lean pork cubes. Cook as above, increasing cooking time to 30 minutes or until meat is tender. Replace 1/2 cup stock with white wine, if desired. Tarragon or thyme may be added to the sauce for flavor.

Chicken Véronique

4 chicken breasts
Salt
Freshly ground pepper
2 tablespoons butter or margarine
1 tablespoon vegetable oil
3 tablespoons all-purpose flour
1/2 cup white wine
1/2 cup chicken stock
Grated peel of 1/2 lemon
1 tablespoon lemon juice
1 bay leaf
1/2 cup whipping cream
1 egg yolk
4 oz. green grapes, halved, seeded

To garnish:
Green grapes or kiwifruit slices
Watercress

1. Preheat oven to 350F (175C).
2. Sprinkle chicken lightly with salt and pepper.
3. Heat butter or margarine and oil in a large skillet. Add chicken; sauté until lightly browned. With tongs, transfer browned chicken to a shallow casserole.
4. Stir flour into fat remaining in skillet; cook 2 minutes, stirring constantly. Stir in wine and stock; bring to a boil. Stir in lemon peel and juice. Season sauce with salt and pepper; pour over chicken. Add bay leaf.
5. Cover casserole with foil or lid. Bake in preheated oven 30 minutes. Place chicken breasts in a warmed serving dish.
6. In a medium saucepan, blend cream with egg yolk; gradually stir in sauce from casserole and grapes. Simmer over low heat 3 to 4 minutes, but do not boil.
7. Discard bay leaf; spoon sauce over chicken. Garnish with grapes or kiwifruit slices and watercress. Makes 4 servings.

Lemon Chicken with Walnuts

1 (2-1/2- to 3-lb.) broiler-fryer chicken, quartered
Salt
Freshly ground pepper
3 tablespoons butter or margarine
2 tablespoons chopped green onion
1 tablespoon all-purpose flour
1 teaspoon ground ginger
1-1/2 cups beef stock
1 tablespoon molasses
Grated peel of 1 lemon
2 tablespoons lemon juice
1/2 cup chopped walnuts

To garnish:
Julienned lemon peel
Walnut halves
Parsley sprigs

1. Preheat oven to 350F (175C).
2. Season chicken with salt and pepper. Melt butter or margarine in a large skillet over medium heat. Add chicken; sauté until browned. With tongs, transfer browned chicken to a large casserole.
3. Add onion to fat remaining in skillet; sauté until soft. Stir in flour and ginger; cook 1 minute, stirring constantly. Gradually stir in stock; boil 1 minute.
4. Stir in molasses, lemon peel, lemon juice and chopped walnuts. Pour sauce over chicken; cover casserole with foil or lid.
5. Bake in preheated oven 1 hour or until chicken is done. Spoon any fat from surface of sauce.
6. Garnish with lemon peel, walnut halves and parsley. Makes 4 servings.

Sausage-Stuffed Chicken Rolls

4 boneless chicken breasts
Salt
Freshly ground pepper
6 to 8 oz. smoked pork sausage
2 tablespoons butter or margarine
1 tablespoon oil
1 onion, sliced
3 tablespoons all-purpose flour
1 cup chicken stock
1/4 cup medium-dry sherry
4 oz. button mushrooms, thickly sliced
12 pimento-stuffed green olives, halved

To garnish:
Parsley sprigs

1. Preheat oven to 350F (175C). Remove skin from chicken; pound chicken lightly between 2 sheets of plastic wrap or waxed paper. Season with salt and pepper.
2. Trim ends from pork sausage; cut 4 (3-inch) pieces. Slice remainder of sausage; reserve for garnish. Place a 3-inch piece of sausage on each chicken breast; roll to enclose sausages. Secure with wooden picks.
3. Heat butter or margarine and oil in a large skillet over medium heat. Add chicken rolls; sauté until browned. With tongs, transfer chicken to a shallow casserole.
4. Add onion to fat remaining in skillet; sauté until lightly browned. Stir in flour; cook 1 minute, stirring frequently. Gradually stir in stock and sherry; bring to a boil. Add mushrooms and olives; season with salt and pepper. Pour sauce over chicken.
5. Cover casserole with foil or lid; bake 35 to 40 minutes or until chicken is tender. Garnish with reserved sausage and parsley. Makes 4 servings.

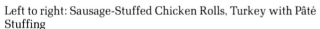

Left to right: Sausage-Stuffed Chicken Rolls, Turkey with Pâté Stuffing

Turkey with Pâté Stuffing

1 (8- to 10-lb.) turkey

Stuffing:
2 tablespoons butter or margarine
1 onion, finely chopped
3 celery stalks, finely chopped
1 garlic clove, crushed, if desired
2 cups fresh breadcrumbs
4 oz. smooth liver pâté
1 teaspoon ground coriander
1 teaspoon Italian seasoning
Salt
Freshly ground pepper

Sauce:
1/2 cup red wine
8 oz. button mushrooms, chopped

Buerre manié:
2 tablespoons butter or margarine
3 tablespoons all-purpose flour

To garnish:
Homemade potato chips
Watercress

1. Debone turkey using illustrations below; do not remove leg and wing bones. Removal of bones makes turkey easier to carve. If desired, turkey can be stuffed without deboning.

2. To make stuffing, melt 2 tablespoons butter or margarine in a medium skillet over medium heat. Add onion, celery and garlic, if desired; sauté 2 to 3 minutes. Put sautéed vegetables into a large bowl; stir in remaining stuffing ingredients.

3. Place deboned turkey breast-side down; spread stuffing inside turkey; close back of turkey over stuffing. Fasten opening with small skewers or sew together with string.

4. Turn turkey over and reshape. Truss loosely. Place stuffed turkey on a roasting rack in a large roasting pan.

5. Roast in preheated oven 3-1/2 to 4 hours or until a thermometer inserted in inner thigh area reaches 180F (85C). Baste occasionally with pan drippings while roasting.

6. Remove turkey from roasting pan. Cover with foil to keep warm; set aside. Degrease pan drippings; add wine and mushrooms to roasting pan. Bring to a boil. In a small bowl, cream butter or margarine and flour together to make buerre manié, then whisk into sauce a little at a time. Boil 3 to 4 minutes, stirring frequently.

7. Place turkey on a serving platter; garnish with potato chips and watercress. Serve sauce separately in a gravy boat or small pitcher. Makes 8 servings.

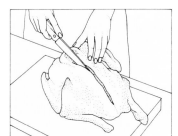

1/Remove tail. Cut along backbone with sharp knife.

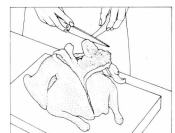

2/Cut flesh and skin from ribs and first wing joint. Remove wing bone.

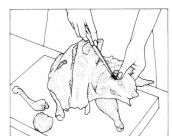

3/Remove flesh and skin from back to thigh.

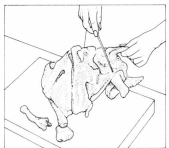

4/Remove flesh from thigh bone. Cut through joint; remove bone.

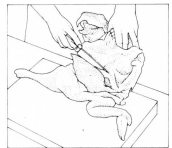

5/Carefully scrape flesh from one side of breast bone.

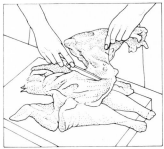

6/Remove flesh from other side of breast bone. Remove carcass.

Duck with Cumberland Sauce

1 (5- to 6-lb.) duck, thawed, if frozen
Salt
Freshly ground pepper
Peel of 1 orange and 1 lemon
Juice of 2 oranges
Juice of 1/2 to 1 lemon
1/4 cup port
3 tablespoons red-currant jelly
2 tablespoons cornstarch
Water

To garnish:
Orange slices
Watercress

1. Preheat oven to 350F (175C).
2. Remove giblets and neck from duck; reserve for other use. Remove excess fat from duck and discard. Prick skin all over to allow fat to escape during roasting. Place duck, breast-side up, on a rack in a shallow roasting pan. Sprinkle duck with salt and pepper.
3. Roast duck in preheated oven 1 hour.
4. Meanwhile, cut fruit peels into julienne strips. Cook in boiling water 5 minutes; drain.
5. In a medium saucepan, heat fruit juices, port and jelly until jelly melts. Add two-thirds of cooked peel. Season with salt and pepper.
6. Spoon off fat from roasting pan, leaving pan juices. Baste duck with orange mixture. Roast duck 1 to 1-1/2 hours longer or until duck is done; baste occasionally with orange mixture. Juices should run clear when duck is pierced between breast and thigh.
7. Place roasted duck on a serving platter; cover with foil to keep warm. Strain pan juices into a saucepan. In a small bowl, combine cornstarch with a little water; stir cornstarch mixture into pan juices. Cook over medium heat until mixture boils and thickens. Serve sauce in a gravy boat or small pitcher.
8. Sprinkle duck with reserved peel; garnish with orange slices and watercress. Makes 4 servings.

The easiest way to remove peel from citrus without white pith is with a vegetable peeler; use as if peeling a potato. If a vegetable peeler is not available, use a small sharp knife to remove peel. Place on a flat surface, peel-side down, and cut off all white pith. Julienne strips are very narrow strips which can be easily cut with a large sharp knife.

Stuffed Turkey Rolls

Stuffing:
1/4 cup butter or margarine
1/2 onion, finely chopped
1 celery stalk, finely chopped
1 cup fresh breadcrumbs
Grated peel of 1/2 orange
8 pimento-stuffed green olives, chopped

Turkey Rolls:
4 uncooked turkey-breast fillets
Salt
Freshly ground pepper
2 tablespoons butter or margarine
1 onion, thinly sliced
1 garlic clove, crushed
About 1/2 cup chicken stock
Juice of 1 orange
2 teaspoons all-purpose flour
Grated peel of 1/2 orange
1 (10-oz.) can cream-style corn

To garnish:
Orange slices
Pimento-stuffed green olives
Watercress

1. Preheat oven to 350F (175C).
2. For stuffing, melt butter or margarine in a heavy medium saucepan over medium heat. Add onion and celery; sauté until soft. Remove from heat; stir in breadcrumbs, orange peel, olives, salt and pepper.
3. Pound turkey fillets between 2 sheets of plastic wrap or waxed paper until 1/4 inch thick. Sprinkle lightly with salt and pepper.
4. Divide stuffing among beaten fillets; roll to enclose stuffing. Secure with wooden picks.
5. Melt butter or margarine in a large skillet over medium heat. Add turkey rolls; sauté until golden brown. With tongs, transfer browned rolls to a shallow casserole.
6. Add onion and garlic to fat remaining in skillet; sauté until golden brown.
7. Add enough stock to orange juice to make 1 cup.
8. Stir flour into onion mixture; cook 1 minute, stirring constantly. Gradually stir in juice and stock; bring to a boil. Stir in remaining orange peel and corn; bring back to a boil. Season sauce with salt and pepper; pour over turkey.
9. Cover casserole with foil or lid; bake in preheated oven 30 minutes.
10. Garnish with orange slices, olives and watercress. Makes 4 servings.

Top to bottom: Duck with Cumberland Sauce, Stuffed Turkey Rolls

Clockwise from top: Duck with Olives, Venison with Juniper Berries, Chicken Breasts with Apples

Chicken Breasts with Apples

4 boneless chicken breasts
Salt
Freshly ground pepper
2 tablespoons butter or margarine
1 tablespoon vegetable oil
1 onion, chopped
1 tablespoon all-purpose flour
1 cup chicken stock
3 Golden Delicious apples, peeled, sliced

To garnish:
4 crisp-cooked bacon slices, crumbled

For those who prefer dark meat, substitute 8 chicken thighs.

1. Preheat oven to 350F (175C).
2. Season chicken with salt and pepper. Heat butter or margarine and oil in a large skillet over medium heat. Add chicken; sauté until browned. With tongs, transfer to a shallow casserole.
3. Add onion to fat remaining in skillet; sauté until lightly colored.
4. Stir flour into skillet; cook 1 minute, stirring constantly. Stir in stock; bring to a boil.
5. Add apples; season with salt and pepper. Pour sauce over chicken.
6. Cover casserole with foil or lid. Bake in preheated oven 35 minutes or until chicken is tender. Garnish with crumbled bacon. Makes 4 servings.

Duck with Olives

2 duck breasts, halved
Salt
Freshly ground pepper
1 tablespoon vegetable oil
1 onion, thinly sliced
1 tablespoon all-purpose flour
1/2 cup chicken stock
6 tablespoons medium-dry sherry
2 tablespoons lemon juice
Grated peel of 1/2 lemon
12 to 16 pimento-stuffed green olives, halved

To garnish:
Lemon slices or wedges
Parsley sprigs, if desired

1. Preheat oven to 350F (175C).
2. Remove any excess fat from duck. Prick skin all over to allow fat to escape during cooking. Sprinkle duck with salt and pepper.
3. Heat oil in a large skillet over medium heat. Add duck; sauté until well browned. Transfer browned duck to a large casserole.
4. Discard all but 1 tablespoon fat from skillet. Add onion; sauté until soft. Stir in flour; cook 1 minute, stirring constantly.
5. Gradually stir in stock, sherry and lemon juice; boil 2 minutes. Add lemon peel and olives. Season sauce with salt and pepper; pour over duck.
6. Cover casserole with foil or lid; bake in preheated oven 40 to 50 minutes or until duck is tender, basting once during cooking.
7. Spoon excess fat from surface of casserole; garnish with lemon and parsley, if desired. Makes 4 servings.

Variation
A whole duck can also, be used if it is first cut into quarters. Use poultry shears, sharp kitchen scissors or a sharp knife to quarter duck.

Venison with Juniper Berries

4 (6-oz.) venison steaks
Salt
Freshly ground pepper
2 tablespoons butter or margarine
Juice and grated peel of 1/2 lemon
12 juniper berries, crushed
1 cup beef stock
1 tablespoon wine vinegar
3 tablespoons whipping cream
1-1/2 teaspoons cornstarch

To garnish:
Carrot and celery sticks
Watercress, if desired

1. Preheat oven to 400F (205C).
2. Sprinkle venison with salt and pepper.
3. Melt butter or margarine in a large skillet. Add venison; sauté until browned. Transfer to a large shallow casserole.
4. Add lemon juice and peel, juniper berries, stock and vinegar to skillet. Boil rapidly 2 to 3 minutes, scraping any browned bits from skillet into sauce. Season with salt and pepper.
5. Pour sauce over venison; cover casserole with foil or lid. Bake in preheated oven 40 minutes or until venison is tender.
6. Drain sauce into a medium saucepan. Keep venison covered and warm. In a small bowl, blend cream with cornstarch; stir some sauce into cornstarch mixture. Stir mixture into pan; simmer 2 to 3 minutes.
7. Arrange venison in a deep serving dish; add sauce. Garnish with carrot and celery sticks and watercress, if desired.

Variation
Substitute 4 (1-inch-thick) pieces of beef-round steak for venison.

Country Roast Chicken

1 (4-lb.) roasting chicken
Salt
Freshly ground black pepper
2 onions, thickly sliced
1 garlic clove, crushed
1 red bell pepper, sliced
1 (16-oz.) can tomatoes
1/2 cup white wine or chicken stock
1 bay leaf
3 oz. sliced salami

To garnish:
Parsley sprigs

1. Preheat oven to 350F (175C).
2. Remove giblets and neck. Chop liver; reserve remaining giblets for other use. Season chicken with salt and pepper.
3. Place chicken in a large heavy casserole. Arrange onions, garlic, red pepper, chopped liver, tomatoes and wine or stock around chicken. Add bay leaf.
4. Chop 2 ounces of salami; reserve remainder for garnish. Add chopped salami to chicken. Cover casserole tightly with lid or foil.
5. Bake in preheated oven 1-1/2 hours or until chicken is tender.
6. Discard bay leaf; garnish with reserved salami and parsley. Makes 4 to 6 servings.

Variation

Baked chicken with salami, peppers, tomatoes and garlic is a dish from the French countryside. Once it is prepared and in the oven, it can be forgotten until ready to serve. It goes well with boiled or creamed potatoes or cooked rice and a green vegetable or salad. Any leftover chicken can be served cold; the flavor is superb. To serve more people, a larger chicken or small turkey, 7 to 8 pounds, can be used; simply increase cooking time to 2-1/2 hours or until chicken or turkey is tender. If you do not have a large enough casserole, use a roasting pan covered tightly with foil. If desired, salami can be omitted or substitute crumbled crisp-cooked bacon.

Cornish Hens with Cherries

2 Cornish hens, thawed, if frozen
Salt
Freshly ground pepper
2 tablespoons vegetable oil
6 oz. small white onions
2 tablespoons all-purpose flour
1 (16-oz.) can dark sweet cherries
About 1/2 cup beef stock
2 tablespoons wine vinegar
2 tablespoons brandy
1 teaspoon dried leaf thyme

Fleurons:
4 oz. puff pastry or 2 to 3 frozen patty shells, thawed
Milk or beaten egg

1. Halve Cornish hens; remove backbone, using a pair of poultry shears, kitchen scissors or a sharp knife. Sprinkle with salt and pepper.
2. Heat oil in a large skillet over medium heat. Add hen halves; sauté until browned. With tongs, transfer to a large casserole.
3. Add onions to fat remaining in skillet; sauté until lightly browned. Stir in flour; cook 1 minute, stirring constantly.
4. Drain juice from cherries; reserve juice. Add enough stock to cherry juice to make 1-1/2 cups. Gradually stir stock mixture into flour; bring to a boil. Add cherries. Season with salt and pepper.
5. Add vinegar, brandy and thyme to sauce; pour over hens. Cover casserole with foil or lid.
6. Bake in preheated oven 1 hour or until hens are tender.
7. To make pastry fleurons, preheat oven to 400F (205C). If baking at same time as Cornish hens, remove casserole when hens are done; increase oven heat to 400F (205C). Roll out puff pastry or patty shells on a lightly floured surface to 1/4 inch thick. Using a fluted pastry cutter, cut into 2-1/2-inch crescents. Place crescents on a ungreased baking sheet; brush with milk or beaten egg. Bake in preheated oven 10 to 15 minutes or until golden brown. Crescents can be made ahead; reheat when needed.
8. Garnish with baked fleurons. Makes 4 servings.

Island Chicken

1 (2-1/2- to 3-lb.) broiler-fryer chicken, quartered
Salt
Freshly ground pepper
2 tablespoons butter or margarine
1 tablespoon vegetable oil
1 onion, sliced
1 garlic clove, crushed
2 celery stalks, sliced
2 tablespoons all-purpose flour
1 cup chicken stock
1/2 cup pineapple juice
Few strands of saffron or 1/4 teaspoon turmeric
1/2 teaspoon ground coriander
1/4 cup shredded coconut

To garnish:
Toasted shredded coconut
Parsley sprigs

1. Preheat oven to 350F (175C).
2. Season chicken with salt and pepper. Heat butter or margarine and oil in a large skillet over medium heat. Add chicken; sauté until well browned. With tongs, transfer browned chicken to a casserole.
3. Add onion, garlic and celery to fat remaining in skillet; sauté until soft. Stir in flour; cook 1 minute, stirring constantly.
4. Stir in stock and pineapple juice; bring to a boil. Stir in saffron or turmeric, coriander and coconut. Season sauce with salt and pepper; pour over chicken.
5. Cover casserole with foil or lid; bake in preheated oven 1 hour or until chicken is tender.
6. Spoon any excess fat from surface of casserole; garnish with toasted coconut and parsley. Makes 4 servings.

Variation
If fresh coconut is available, replace shredded coconut with 2 ounces freshly grated coconut. If desired, 1/2 cup coconut milk can replace 1/2 cup stock.

Clockwise from left: Country Roast Chicken, Cornish Hens with Cherries, Island Chicken

Festive Pheasant with Chestnuts

2 (2- to 3-lb.) pheasants, cut up
Salt
Freshly ground pepper
2 tablespoons butter or margarine
1 tablespoon vegetable oil
8 to 12 oz. chestnuts, roasted or boiled, peeled
1 bay leaf
8 oz. small white onions
2 tablespoons all-purpose flour
1 cup dry white wine
1/2 cup chicken stock
1 tablespoon red-currant jelly
Julienned peel of 1 orange
Juice of 2 oranges
2 tablespoons brandy, if desired

To garnish:
Orange slices
Watercress

1. Preheat oven to 350F (175C).
2. Sprinkle pheasants with salt and pepper. Heat butter or margarine and oil in a large skillet over medium heat. Add pheasant pieces; sauté until browned. With tongs, transfer browned pheasant to a large casserole. Arrange chestnuts around pheasant. Add bay leaf.

3. Add onions to fat remaining in skillet; sauté until lightly colored. Use a slotted spoon to place sautéed onions over pheasants and chestnuts.
4. Stir flour into fat remaining in skillet; cook 1 minute, stirring constantly. Gradually stir in wine and stock; bring to a boil. Stir in jelly, orange peel and juice and brandy, if desired. Season with salt and pepper; pour sauce over pheasants.
5. Cover casserole tightly with foil or lid; bake in preheated oven 50 to 60 minutes or until pheasants are tender.
6. Discard bay leaf; serve pheasant pieces on a platter surrounded by onions and chestnuts. Spoon some sauce over pheasants; serve remaining sauce in a gravy boat or small pitcher. Garnish with orange slices and watercress. Makes 4 servings.

Variation
Cornish hens are a good substitute if pheasants are not available. Canned chestnuts can be substituted for fresh ones; drain and rinse with cold water before using.

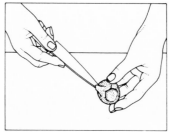
1/Pierce each chestnut; boil 25 minutes.

2/Use a sharp knife to peel.

Festive Pheasant with Chestnuts

Cornish Hens Véronique

4 Cornish hens, thawed, if frozen
Salt
Freshly ground pepper
4 bacon slices
1 cup beef stock
2 tablespoons brandy
Juice of 1 orange
2 tablespoons medium-dry sherry
1 tablespoon cornstarch
4 to 6 oz. green grapes, halved, seeded

To garnish:
Orange slices, cut in half
Green grapes

1. Preheat oven to 350F (175C).
2. Sprinkle hens with salt and pepper. Place seasoned hens in a roasting pan. Cut bacon slices in half; lay 2 halves across each hen.
3. Roast in preheated oven 1-1/2 hours or until hens are tender.
4. Place roasted Cornish hens on a serving platter; cover with foil to keep warm.
5. Remove fat from pan drippings; deglaze pan with stock. Add brandy and orange juice. In a small bowl, combine sherry and cornstarch. Stir cornstarch mixture into pan. Bring to a boil, stirring constantly; add halved grapes. Simmer sauce 3 to 4 minutes; spoon over Cornish hens.
6. Garnish with orange slices and grapes. Makes 6 to 8 servings.

Cornish Hens Véronique

Shrimp Veracruz-Style
Photo on cover

1 lb. uncooked medium shrimp
1 large green pepper
3 tablespoons vegetable oil
1 small onion, chopped
5 small tomatoes, peeled, chopped (1-1/4 lbs.)
12 pimento-stuffed green olives
1-1/2 teaspoons capers
1 bay leaf
1/2 teaspoon sugar
1/2 teaspoon salt
Lime juice

1. Peel shrimp; devein. Set shrimp aside.
2. Cut green pepper into 1-1/2" x 1/2" strips. Heat 1 tablespoon oil in a large saucepan over medium heat. Add onion and pepper strips. Cook until onion is tender but not browned.
3. Add tomatoes, olives, capers, bay leaf, sugar and salt. Bring to a boil; reduce heat. Cover; simmer 20 minutes. Remove bay leaf from tomato sauce.
4. Heat remaining oil in a large skillet over medium heat. Add peeled shrimp. Cook about 3 minutes or until shrimp turn pink. Sprinkle a few drops of lime juice over shrimp.
5. Add cooked tomato sauce to shrimp. Cook 3 to 4 minutes, stirring constantly.
6. Serve hot shrimp mixture immediately. Makes 4 servings.

Squid Provençal

3 lbs. fresh squid
2 tablespoons olive oil
2 onions, sliced
1 garlic clove, crushed
2 tablespoons all-purpose flour
1 tablespoon tomato paste
1 cup dry white wine or vermouth
1/2 cup chicken stock
1 tablespoon lemon juice
3 small tomatoes, peeled, sliced
Salt
Freshly ground black pepper
1/2 teaspoon paprika
Pinch of red (cayenne) pepper
Pinch of sugar
1 bay leaf

To garnish:
Chopped fresh parsley

1. Preheat oven to 325F (165C).
2. To prepare squid, pull gently to separate head from body and tentacles. Pull transparent pen from body. Remove speckled membrane, if desired. Wash cleaned squid; cut into rings. Remove beak from base of tentacles; cut tentacles into pieces. Discard head and ink sac.
3. Heat oil in a flameproof casserole. Add onions and garlic; sauté until lightly browned. Add squid; sauté until lightly colored.
4. Stir in flour; cook 2 minutes, stirring constantly. Stir in tomato paste, wine or vermouth, stock and lemon juice. Bring to a boil; add tomatoes, salt, black pepper, paprika, red pepper and sugar to taste. Add bay leaf; cover casserole with foil or lid.
5. Bake in preheated oven 45 minutes or until squid is tender. Discard bay leaf. Garnish with parsley. Makes 4 servings.

1/Pull head from body. 2/Remove pen and discard.

Halibut Catalan

4 halibut fillets or steaks
6 tablespoons olive oil
Juice of 1 lemon
1 small onion, finely chopped
1 garlic clove, crushed
1 tablespoon all-purpose flour
8 oz. tomatoes, peeled, seeded, chopped
1 tablespoon tomato paste
1 cup dry white wine
Salt
Freshly ground pepper

To garnish:
1/2 cup finely chopped nuts
2 tablespoons chopped fresh parsley

Around the Mediterranean, people love to cook white fish in thick sauces, highly flavored with tomatoes, onion and garlic. Halibut lends itself well to this flavor combination.

1. Place fish in a large glass bowl or pan. Combine 1/4 cup oil and lemon juice; pour over fish. Cover and refrigerate 1 to 2 hours, turning occasionally.
2. Preheat oven to 350F (175C).
3. Heat remaining 2 tablespoons oil in a deep flameproof casserole. Add onion and garlic; sauté until golden brown. Stir flour into onions and garlic; cook 2 minutes, stirring constantly. Stir in tomatoes, tomato paste and wine. Slowly bring to a boil. Season with salt and pepper.
4. Add marinated fish to sauce; turn to coat.
5. Cover casserole with foil or lid; bake in preheated oven 10 to 15 minutes or until fish tests done.
6. Garnish casserole with nuts and parsley. Makes 4 servings.

Halibut Catalan

Baked Haddock with Wine Sauce

1-1/2 lbs. haddock fillets, skinned
Salt
Freshly ground pepper
3 tomatoes, peeled, quartered, seeded
6 oz. mushrooms or apples, sliced
2 tablespoons butter or margarine
2 tablespoons all-purpose flour
1 cup white wine or apple juice
1 tablespoon chopped fresh chives or 1 teaspoon dried
   chives
1/4 teaspoon grated lemon peel
1 tablespoon lemon juice

1. Preheat oven to 350F (175C). Grease a shallow casserole; set aside.
2. Cut fish into 4 equal portions, sprinkle lightly with salt and pepper. Roll fillets; place in greased casserole. Arrange tomatoes and mushrooms or apples over fish.
3. Melt butter or margarine in a small saucepan. Stir in flour; cook 1 minute, stirring constantly. Gradually stir in wine or juice; bring to a boil. Add chives, lemon peel and juice. Season with salt and pepper.
4. Pour sauce over fish; cover casserole with foil or lid.
5. Bake in preheated oven 20 minutes or until fish tests done. Place fish on a serving dish; reduce sauce, if necessary, before pouring over fish. Makes 4 servings.

Bouillabaisse

2-1/2 lbs. mixed fish and shellfish
6 tablespoons vegetable oil
2 large onions, thinly sliced
2 celery stalks, thinly sliced
2 carrots, diced
2 garlic cloves, crushed
1 lb. tomatoes, peeled, sliced
1 bouquet garni
Grated peel of 1/2 lemon
Juice of 1 lemon
Pinch of saffron or turmeric
Salt
Freshly ground pepper
About 1 cup white wine

To garnish:
Few whole shrimp or crayfish
Chopped fresh parsley, if desired

A true bouillabaisse is made from at least 8 types of fish and shellfish, but many of the authentic types are only available in Mediterranean areas. Your fish market should help with a good selection if you explain your requirements. If you prefer more liquid in your stew, add 1 to 2 cups fish stock or chicken stock with the wine.

Clockwise from top left: Baked Haddock with Wine Sauce, Salmon & Asparagus, Scallops & Mushrooms in Cream Sauce, Bouillabaisse

1. Clean fish, removing skin and any loose bones. Rinse fish in cold water; cut into 2-inch pieces. Peel all shrimp or crayfish except those used for garnish; rinse in cold water.
2. Heat oil in a large heavy saucepan over medium heat. Add onions, celery, carrots and garlic; sauté until soft but only lightly colored.
3. Add tomatoes, bouquet garni, lemon peel and juice and saffron or turmeric to saucepan; season vegetable mixture with salt and pepper. Bring to a boil. Reduce heat; simmer 10 minutes.
4. Lay fish over vegetables; add enough wine to almost cover fish. Simmer gently 2 to 3 minutes.
5. Add shellfish to saucepan; simmer an additional 4 to 6 minutes longer or until fish and shellfish are done. Do not overcook. Discard any mussels that do not open. Discard bouquet garni; adjust seasoning, if necessary.
6. Top with whole shrimp or crayfish and parsley, if desired. Serve with crusty bread. Makes 4 servings.

When fresh scallops are unavailable, use frozen ones. Be sure they are completely thawed before using.

1. If scallops are large, cut into 2 or 3 pieces.
2. Heat butter or margarine in a large skillet over medium heat. Add onion, celery and mushrooms; sauté until soft but not colored. Add scallops to vegetable mixture; sauté 1 to 2 minutes. Use a slotted spoon to remove scallops and vegetables from skillet; place in a flameproof casserole.
3. Stir flour into fat remaining in skillet; cook 2 minutes, stirring constantly. Gradually stir in wine, stock or water and lemon juice; bring to a boil. Simmer until slightly thickened and smooth, stirring frequently. Season with hot-pepper sauce, salt and pepper. Stir whipping cream or sour cream into sauce; pour over scallops and vegetables.
4. Preheat broiler.
5. Combine breadcrumbs and cheese; sprinkle over scallop mixture. Place casserole under preheated broiler until top is crisp and brown. Serve at once, garnished with cucumber and lemon slices. Makes 4 servings.

Salmon & Asparagus

4 (1-inch-thick) salmon steaks
Salt
Freshly ground pepper
1 (12-oz.) can asparagus spears
2 teaspoons cornstarch
6 tablespoons whipping cream
2 teaspoons lemon juice

To garnish:
Lemon wedges

1. Preheat oven to 375F (190C). Grease a shallow baking pan large enough to hold salmon in a single layer.
2. Place salmon in greased baking pan; sprinkle with salt and pepper.
3. Bake salmon in preheated oven 10 minutes or until fish tests done.
4. Meanwhile, make sauce. Drain asparagus, reserving 6 tablespoons of liquid. Reserve 12 asparagus spears for garnish; coarsely chop remainder. Set chopped asparagus aside.
5. In a small saucepan, blend cornstarch with cream. Stir in reserved asparagus liquid and lemon juice. Cook mixture over low heat, stirring constantly, until slightly thickened and smooth. Season with salt and pepper. Gently stir in chopped asparagus; cook until hot.
6. Arrange baked salmon on a warmed serving plate. Spoon hot sauce over salmon. Garnish with reserved asparagus spears and lemon wedges. Makes 4 servings.

Scallops & Mushrooms in Cream Sauce

10 to 12 sea scallops
3 tablespoons butter or margarine
1 large onion, finely chopped
2 to 3 celery stalks, thinly sliced
6 oz. button mushrooms, quartered or sliced
2 tablespoons all-purpose flour
1/2 cup white wine
1/2 cup chicken stock or water
1 tablespoon lemon juice
Few drops of hot-pepper sauce
Salt
Freshly ground black pepper
3 tablespoons whipping cream or dairy sour cream

Crumb topping:
1 cup fresh white breadcrumbs
1/4 cup grated Cheddar cheese (1 oz.)

To garnish:
Cucumber slices
Lemon slices

Sole Paupiettes with Smoked Salmon

8 (3- to 4-oz.) sole fillets
Salt
Freshly ground pepper
4 oz. smoked salmon
3 tablespoons butter or margarine
1/4 cup all-purpose flour
1/2 cup dry white wine
1 cup milk
3 tablespoons half and half
1 tablespoon lemon juice

To garnish:
8 whole shrimp
Fresh parsley, fennel or dill, if desired

1. Preheat oven to 400F (205C). Grease a shallow baking dish.
2. Wipe sole fillets with a damp paper towel; sprinkle with salt and pepper. Divide smoked salmon into 8 pieces; lay 1 piece on skin-side of each fillet. Roll up loosely towards tail; place rolls in greased dish.
3. Melt butter or margarine in a medium saucepan. Stir in flour; cook 1 minute, stirring constantly. Gradually stir in wine and milk; boil 2 minutes. Season with salt and pepper; stir in half and half and lemon juice. Pour sauce over fish.
4. Cover dish with foil or lid; bake in preheated oven 10 minutes or until fish tests done. Do not overcook.
5. Garnish sole paupiettes with whole shrimp and parsley, fennel or dill, if desired. Makes 4 servings.

Sole fillets can be used with the skins on. If you prefer to remove them, it is simple to do. Place fillet on a wooden board or work surface, skin-side down. Using a sharp knife and beginning at the tail, carefully run the knife along towards the head, keeping blade slanting downwards, working away from you. The fillet will then lift away easily from skin. If you work from head to tail, flesh will crumble, break and not come away from skin in one piece. If the fish slips, sprinkle a little coarse salt on work surface before you begin; rinse fillets before cooking to remove any salt.

Sole & Shrimp with Whiskey

4 (3- to 4-oz.) sole or flounder fillets
1/2 cup water or fish stock
1/2 cup white wine
4 peppercorns
1 bay leaf
Parsley sprig
Salt

Shrimp topping:
3 tablespoons butter or margarine
1 small onion, halved, sliced
1 large tomato, peeled, seeded, sliced
1/2 lb. peeled, deveined small shrimp
Salt
Freshly ground white pepper
2 to 3 tablespoons whiskey

To garnish:
Fresh dill, if desired

1. Preheat oven to 350F (175C). Grease a shallow baking dish.
2. Place fillets in a single layer in greased baking dish.
3. In a small saucepan, combine water or fish stock, wine, peppercorns, bay leaf, parsley and salt. Bring to a boil; pour over fillets. Cover fillets with a sheet of buttered waxed paper.
4. Poach in preheated oven 10 minutes or until fish tests done.
5. Meanwhile, make shrimp topping. Melt butter or margarine in a large skillet over medium heat. Add onion; sauté 3 to 4 minutes. Add tomatoes and shrimp; sauté 2 to 3 minutes or until shrimp turn pink, stirring frequently. Season with salt and white pepper. Pour whiskey into skillet. Heat mixture; ignite carefully.
6. Place poached fillets on a serving plate; discard poaching liquid or reserve for other use. When flame dies, spoon shrimp mixture over fillets; garnish with dill, if desired. Makes 4 servings.

Turbot with Champagne

1/4 cup butter or margarine
3 shallots or 1 medium onion, finely chopped
1 bunch watercress
1 cup milk
Salt
Freshly ground white pepper
4 (4- to 6-oz.) turbot or other white-fish steaks
1 bay leaf
1 cup champagne

1. Preheat oven to 350F (175C). Grease a baking dish large enough to hold the fish in a single layer.
2. Melt butter or margarine in a heavy medium saucepan over medium heat. Add shallots or onion; sauté until soft. Reserve a few sprigs of watercress for garnish; finely chop remaining watercress. Add chopped watercress to skillet; sauté 2 minutes, stirring constantly.
3. Add milk; simmer 15 minutes. Season with salt and white pepper; allow to cool slightly.
4. While watercress mixture is simmering, bake fish. Place fish in a single layer in greased baking dish. Add bay leaf and champagne; season with salt and pepper. Cover baking dish with foil or lid; bake in preheated oven 10 minutes or until fish tests done.
5. Place fish on a warmed serving platter. Cover with foil to keep warm. Discard bay leaf.
6. In a blender or food processor fitted with a steel blade, puree cooking liquid and watercress mixture. Pour some sauce over fish; serve remaining sauce separately. Garnish fish with reserved watercress. Makes 4 servings.

Clockwise from top left: Mussel Kabobs, Turbot with Champagne, Sole & Shrimp with Whiskey

Variation
Any dry white wine can be substituted for the champagne. If turbot is not available, substitute halibut or another white fish.

Mussel Kabobs

1/2 cup butter or margarine, melted
1/4 cup white wine
1 garlic clove, crushed
3 to 4 drops hot-pepper sauce
Salt
Freshly ground white pepper
24 mussels, cleaned, shucked
1 large green pepper, cut into 1-inch squares
1 large red bell pepper, cut into 1-inch squares
About 8 oz. ready-to-eat ham, cut in 1-inch chunks

1. Soak bamboo skewers in water 30 minutes. Preheat grill or broiler.
2. In a small saucepan over medium heat, combine butter or margarine, wine, garlic and hot-pepper sauce. Season mixture with salt and white pepper. Heat until hot.
3. Thread mussels, green and red peppers and ham on skewers, as shown below. Baste each kabob with butter mixture.
4. Place skewers on rack 4 inches from coals or high heat. Cook 2 minutes; baste with butter sauce and turn. Cook 2 minutes longer, basting once. Reheat sauce; serve separately. Makes 4 servings.

Rainbow Trout with Brown Butter & Almonds

4 rainbow trout, cleaned
Salt
Freshly ground pepper
1/4 cup butter or margarine
1/2 cup coarsely chopped blanched almonds
2 tablespoons lemon juice

Stuffing balls:
1-1/2 cups fresh breadcrumbs
2 tablespoons grated onion
Grated peel of 1/2 lemon
1/4 cup finely chopped blanched almonds
1 tablespoon chopped fresh parsley
1 egg yolk

To garnish:
Parsley sprigs
Lemon wedges

1. Preheat oven to 400F (205C). Lightly grease a shallow baking dish. Season fish with salt and pepper. Place fish head-to-tail in a single layer in greased baking dish.
2. Melt butter or margarine in a small skillet. Add coarsely chopped almonds; sauté until almonds and butter or margarine are lightly browned. Remove skillet from heat. Stir in lemon juice; quickly pour mixture over fish.
3. For stuffing balls, in a medium bowl, combine breadcrumbs, onion, lemon peel, finely chopped almonds and parsley. Season with salt and pepper; stir in egg yolk.
4. Shape stuffing mixture into 8 balls; arrange around fish.
5. Cover dish with foil or lid; bake in preheated oven 20 minutes or until fish tests done. Do not overcook. Serve garnished with parsley and lemon wedges. Makes 4 servings.

Moules Marinière

4 quarts fresh mussels
1/4 cup butter or margarine
2 onions, finely chopped
2 carrots, finely chopped
1 to 2 garlic cloves, crushed
1-1/2 cups white wine or apple juice
1 tablespoon lemon juice
Salt
Freshly ground pepper
2 bay leaves
4 teaspoons cornstarch
3 tablespoons whipping cream

To garnish:
Chopped fresh parsley

1. Scrub mussels with a brush to remove all dirt. Discard any that are broken or do not close when given a sharp tap. Pull away beards.
2. Melt butter or margarine in a large kettle. Add onions, carrots and garlic; sauté gently until soft but not browned.
3. Add wine or apple juice and lemon juice to kettle; bring mixture to a boil. Add salt, pepper and bay leaves.
4. Add mussels to kettle; stir well. Cover kettle tightly with foil or lid. Simmer 4 to 8 minutes or until mussels open. Discard bay leaves and any mussels that do not open.
5. Ladle mussels into 4 large serving bowls.
6. In a small bowl, blend cornstarch with cream; stir cornstarch mixture into cooking liquid in kettle. Bring slowly to a boil; stir constantly until slightly thickened.
7. Spoon sauce over mussels; sprinkle mussels with parsley. Serve with fresh crusty bread. Makes 4 servings.

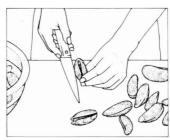 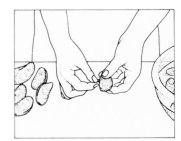

1/Check mussels by tapping to close.

2/Pull away hairy beard.

Seafood Casserole

1-1/2 lbs. haddock or cod fillets, skinned
4 oz. peeled, deveined shrimp
3 to 4 sea scallops, quartered
1 (12-oz.) can whole-kernel corn, drained
1 to 2 canned red pimentos, sliced, if desired
3 tablespoons butter or margarine
1/4 cup all-purpose flour
1 cup milk
1 teaspoon Dijon-style mustard
1 tablespoon lemon juice
1/4 cup dairy sour cream
Salt
Freshly ground pepper

To garnish:
Whole shrimp

1. Preheat oven to 350F (175C).
2. Cut fish into 1-inch cubes. In a large, fairly shallow casserole, combine fish cubes, shrimp, scallops, corn and pimentos, if desired.
3. Melt butter or margarine in a medium saucepan. Stir in flour; cook 1 minute, stirring constantly. Stir in milk; bring to a boil. Remove from heat.
4. Stir mustard, lemon juice and sour cream into sauce; season with salt and pepper. Pour over fish, shellfish and corn.
5. Bake in preheated oven 20 minutes or until fish tests done.
6. Stir casserole gently before serving. Garnish with whole shrimp. Makes 6 servings

Variation

Milk can be replaced by 1/2 cup each of apple juice or white wine and chicken stock. A quick fish stock can be made by simmering fish parts in 1 cup water, 1 sliced onion, 1 chopped carrot, 2 parsley sprigs and a bay leaf. Bring to a boil; then reduce heat and simmer gently 15 minutes. Strain and use in place of chicken stock.

Top to bottom: Moules Marinière, Rainbow Trout with Brown Butter & Almonds, Seafood Casserole

Fish Paella

1/2 cup butter or margarine
1 tablespoon vegetable oil
1 large onion, sliced
1 to 2 garlic cloves, crushed
1 medium red bell pepper, thinly sliced
1 medium green pepper, thinly sliced
1 cup uncooked long-grain white rice
6 oz. mushrooms, thickly sliced
2 bay leaves
2 cups chicken stock or fish stock
Pinch of saffron or 1/2 teaspoon turmeric
Salt
Freshly ground black pepper
2 (4- to 5-oz.) sole fillets, cut into 1-inch pieces
4 to 6 oz. fresh, frozen or canned crabmeat, flaked
4 oz. peeled, deveined shrimp
4 to 8 crayfish
4 oz. frozen green peas, thawed
2 cups fresh mussels, cleaned, if desired

1. Preheat oven to 350F (175C). Grease a paella pan or a shallow 3-quart casserole; set aside.
2. Heat butter or margarine and oil in a medium skillet over medium heat. Add onion and garlic; sauté until soft. Add red and green peppers; cook 3 minutes. Place in greased paella pan or casserole.
3. Add rice, mushrooms and bay leaves.
4. Bring stock to a boil; add saffron or turmeric. Season with salt and pepper; pour over rice mixture.
5. Cover with foil or lid; bake in preheated oven 15 minutes.
6. Stir rice mixture; add sole, crabmeat, shrimp, crayfish, peas and mussels, if desired. Cover and bake 10 minutes longer or until mussels have opened. Discard bay leaves and any mussels that do not open. Makes 4 servings.

Greek-Style Halibut

4 (1-inch-thick) halibut steaks
Salt
Freshly ground pepper
6 oz. button mushrooms, halved, if large
1 bay leaf
2 tablespoons vegetable oil
1 medium onion, chopped
2 carrots, diced
1 garlic clove, crushed
6 tablespoons white wine

To garnish:
Tomato slices
Chopped fresh parsley

1. Preheat oven to 375F (190C). Grease a shallow baking dish large enough to hold fish in a single layer.
2. Lay fish in greased dish; season with salt and pepper. Add mushrooms and bay leaf.
3. Heat oil in a medium skillet over medium heat. Add onion, carrots and garlic; sauté gently until onion and garlic are soft but not browned.
4. Add wine to skillet; boil 1 minute. Pour over fish.
5. Cover dish with foil or lid; bake in preheated oven 10 to 15 minutes or until fish tests done.
6. Discard bay leaf; garnish with tomato and parsley. Makes 4 servings.

Scallop & Artichoke Medley

8 oz. Jerusalem artichokes (sunchokes)
Water
2 tablespoons lemon juice
1/4 cup butter or margarine
1 large onion, sliced
1 bay leaf
1/2 cup dry vermouth
1/2 cup water
12 sea scallops
Salt
Freshly ground pepper
2 tablespoons cornstarch
3 tablespoons whipping cream or dairy sour cream
1 to 2 canned red pimentos, sliced, if desired

To garnish:
Croutons
Parsley sprigs

1. Peel artichokes; slice into a medium bowl of cold water containing 1 tablespoon lemon juice.
2. Melt butter or margarine in a heavy medium saucepan. Add onion; sauté gently until soft. Drain sliced artichokes. Add to pan; sauté gently 3 to 4 minutes.
3. Add bay leaf, vermouth, remaining 1 tablespoon lemon juice and 1/2 cup water to pan; simmer 10 to 15 minutes or until artichokes are tender.
4. Cut scallops into 3 or 4 pieces; add to saucepan. Season with salt and pepper. Simmer scallops 1 to 2 minutes. Discard bay leaf.
5. Use a slotted spoon to place vegetables and scallops in a deep serving dish; keep warm.
6. In a small bowl, blend cornstarch with cream or sour cream; whisk cornstarch mixture into saucepan. Cook, stirring constantly until sauce is slightly thickened. Add strips of pimento, if desired. Pour sauce over scallops and vegetables.
7. Garnish with croutons and parsley. Makes 4 servings.

Halibut in Crab Sauce

4 (1-inch thick) halibut steaks
Salt
Freshly ground pepper
2 tablespoons butter or margarine
1 tablespoon finely chopped onion
3 tablespoons all-purpose flour
1 cup milk
1 tablespoon lemon juice
1 teaspoon Angostura bitters
4 oz. fresh, frozen or canned crabmeat, flaked

To garnish:
Whole shrimp
Cucumber sticks

Crabmeat in this recipe makes it a dish for special occasions. Any firm white fish can be substituted for halibut.

1. Preheat oven to 350F (175C). Grease a shallow baking dish large enough to hold fish in a single layer.
2. Place fish steaks in greased baking dish; season with salt and pepper.
3. Melt butter or margarine in a skillet. Add onion; sauté gently until soft. Stir in flour; cook 1 minute, stirring constantly. Gradually stir in milk; bring to a boil.
4. Stir in lemon juice, bitters and crabmeat; season with salt and pepper. Pour sauce over fish.
5. Cover with foil or lid; bake in preheated oven 10 minutes or until fish tests done.
6. Arrange baked fish and sauce on a serving dish; garnish with whole shrimp and cucumber sticks. Makes 4 servings.

Clockwise from left: Fish Paella, Halibut in Crab Sauce, Greek-Style Halibut

Grilled Hawaiian Chicken

1 (3- to 3-1/2-lb.) broiler-fryer chicken, cut up
Vegetable oil
Salt
Freshly ground pepper

Pineapple glaze:
1 (20-oz.) can sliced pineapple
2 tablespoons cornstarch
1/2 cup water
1/2 cup packed brown sugar
1/4 teaspoon salt
1/2 teaspoon Worcestershire sauce
1/4 cup white vinegar
2 tablespoons chili sauce
1/3 cup ketchup

---

1. Preheat grill.
2. Brush chicken pieces with oil; season with salt and pepper. Place on grill, skin-side down, 5 to 6 inches from source of heat. Grill over medium-hot coals 35 minutes, turning chicken occasionally with tongs.
3. Prepare pineapple glaze while chicken is cooking. To prepare glaze, drain pineapple juice into a small saucepan; reserve pineapple slices for garnish. In a small bowl, combine cornstarch and water. Gradually stir into pineapple juice. Cook over medium heat, stirring constantly, until smooth and thickened. Add brown sugar, salt, Worcestershire sauce, vinegar, chili sauce and ketchup. Stir well; simmer 3 to 4 minutes, stirring frequently.
4. Brush chicken with glaze; grill 10 to 15 minutes longer or until chicken is tender, turning chicken and brushing often with glaze.
5. Heat remaining glaze; pour into a gravy boat or small pitcher. If desired, grill pineapple slices 2 to 3 minutes on each side. Makes 4 servings.

Leg of Lamb with Rice Stuffing

Stuffing:
2 tablespoons butter or margarine
1 onion, finely chopped
2 celery stalks, finely chopped
3 crisp-cooked bacon slices, crumbled
1/2 cup cooked rice
Pinch of ground allspice
Salt
Freshly ground pepper
1 (8-oz.) can apricot halves
1 cup cooked prunes
1 egg yolk
1 (4-1/2-lb.) lamb-leg roast, boneless
1 tablespoon butter or margarine, melted

Sauce:
1/2 cup dry white wine
1/2 cup beef stock
1-1/2 tablespoons cornstarch

To garnish:
Sliced zucchini, lightly cooked
Fresh rosemary or parsley sprigs

---

1. Preheat oven to 350F (175C).
2. For stuffing, melt butter or margarine in a large skillet over medium heat. Add onion and celery; sauté until lightly browned. Place sautéed mixture in a large bowl; add bacon, rice and allspice. Season with salt and pepper.
3. Drain apricots. Reserve 1/4 cup juice; set aside. Chop 4 apricots and 4 prunes; set aside remaining fruit. Add chopped fruit to stuffing; stir well. Add egg yolk; stir until mixture binds together.
4. Use stuffing to fill bone cavity of lamb; reshape roast, using a trussing needle and fine string or several skewers.
5. Brush surface of meat with melted butter or margarine; sprinkle with salt. Place on a rack in a roasting pan. Roast in preheated oven 2-1/4 to 3 hours or to desired doneness. Baste with pan drippings once or twice during cooking.
6. Remove string or skewers from lamb; place lamb on a serving dish. Deglaze roasting pan with wine; add stock. In a small bowl, combine cornstarch with reserved apricot juice. Stir cornstarch mixture into roasting pan; bring to a boil. Simmer 2 to 3 minutes or until slightly thickened. Season with salt and pepper. Remove fat from surface of sauce. Serve in a gravy boat.
7. Arrange remaining apricots and prunes, zucchini and herbs around roast. Serve with potatoes cooked in their skins, coated with butter or margarine and chopped parsley. Makes 6 to 8 servings.

Leg of Lamb with Rice Stuffing

Flambéed Beef Stew

2 lbs. beef stew cubes
1 garlic clove, crushed
1 tablespoon chopped fresh parsley
1 teaspoon dried leaf thyme
1/2 cup dry red wine
2 tablespoons vegetable oil
1/4 cup brandy
2 onions, sliced
1-1/2 cups beef stock
1 bay leaf
Salt
Freshly ground pepper
2 carrots, sliced

To garnish:
Chopped fresh parsley

---

1. Place beef cubes in a glass or ceramic bowl. Add garlic, parsley, thyme and wine; stir well.
2. Cover and refrigerate about 2 hours, stirring at least once.
3. Drain beef, reserving marinade; pat beef dry with paper towels. Heat oil in a large skillet over medium heat. Add marinated beef cubes; sauté until browned. Warm brandy in a small saucepan; pour over beef cubes. Ignite brandy carefully. When flame dies, remove browned beef cubes to a large saucepan.
4. Add onions, reserved marinade, stock and bay leaf to saucepan. Season with salt and pepper.
5. Cover saucepan; simmer stew over low heat 2-1/2 to 3 hours or until beef is tender. Add carrots to stew after 1 hour.
6. To serve, discard bay leaf; sprinkle heavily with chopped parsley. Makes 5 to 6 servings.

Variation
Chicken or turkey tastes very good when cooked in this way. Allow 1 chicken quarter or 2 smaller pieces of chicken or a turkey thigh per person, with or without bone. Cook for 1-1/4 to 1-1/2 hours or until meat is very tender. White wine may be used in place of red wine.

Chicken Gumbo

3 tablespoons vegetable oil
1 (2-1/2- to 3-lb.) broiler-fryer chicken, cut up
2 tablespoons butter or margarine
1 large onion, sliced
1 garlic clove, crushed
1 green pepper, sliced
1 tablespoon all-purpose flour
1 (16-oz.) can tomatoes
1 cup chicken stock
1 (10-oz.) pkg. frozen okra or 12 oz. fresh okra, trimmed
2 tablespoons tomato paste
2 teaspoons Worcestershire sauce
Pinch of ground cloves
Pinch of chili powder
1/4 teaspoon dried leaf basil
Salt
Freshly ground black pepper
2 cups cooked long-grain white rice

To garnish:
2 tablespoons chopped fresh parsley

Okra is an unusual ingredient which adds flavor and texture. It is used as a thickening agent in casseroles and stews. When available, fresh okra gives better results and color.

1. Preheat oven to 350F (175C).
2. Heat oil in a large skillet over medium heat. Add chicken pieces; sauté until golden brown. Transfer to a 3-quart casserole.
3. Melt butter or margarine in a heavy, medium saucepan. Add onion, garlic and green pepper; sauté 2 minutes or until soft.
4. Stir flour into vegetable mixture; gradually stir in tomatoes and stock. Bring to a boil, stirring frequently.
5. Thickly slice okra; add to pan with tomato paste, Worcestershire sauce, cloves, chili powder, basil, salt and pepper. Simmer gently 5 minutes.
6. Pour okra mixture over chicken. Cover casserole with foil or lid; bake in preheated oven 1 hour or until tender.
7. To serve, spoon gumbo over hot cooked rice; garnish with fresh parsley. Makes 4 servings.

Jambalaya

Jambalaya

2 tablespoons vegetable oil
2 tablespoons butter or margarine
1 lb. lean pork, cut into narrow strips
1 large onion, chopped
1 green pepper, sliced
1 red bell pepper, sliced
4 oz. mushrooms, thickly sliced
1 cup uncooked long-grain white rice
2 cups or more chicken stock
1/4 teaspoon ground allspice
4 oz. smoked sausage, sliced or chopped
Salt
Freshly ground black pepper
4 oz. peeled deveined shrimp

To garnish:
Few whole shrimp
Tomato wedges

1. Preheat oven to 350F (175C).
2. Heat oil and butter or margarine in a large heavy saucepan. Add pork strips; sauté until well browned. With a slotted spoon, transfer browned pork to a 3-quart casserole.
3. Add onion and green and red peppers to fat remaining in skillet; sauté 3 to 4 minutes.
4. Add mushrooms; sauté 1 minute. Stir in rice and 2 cups stock; bring to a boil.
5. Add allspice and smoked sausage; season with salt and pepper. Spoon seasoned rice mixture over pork; stir well. Cover casserole tightly with foil or lid. Bake in preheated oven 50 minutes.
6. Stir well; add shrimp and additional boiling stock, if necessary. Recover casserole; bake 10 minutes longer or until liquid has been absorbed and meat is tender.
7. To serve, garnish jambalaya with whole shrimp and tomato wedges. Makes 4 servings.

Left to right: Steak Bake with Mushrooms & Potatoes, Spicy Sweet & Sour Beef

Steak Bake with Mushrooms & Potatoes

1/3 cup all-purpose flour
Salt
Freshly ground pepper
1-1/2 lbs. beef-round steak, cut into 4 serving pieces
1/4 cup butter or margarine
1 large onion, sliced
6 oz. mushrooms, sliced
3/4 cup beef stock
1 lb. potatoes, peeled, cut into 1-inch cubes (3 cups)
1/2 cup half and half

To garnish:
Chopped fresh parsley
Cooked carrot sticks

1. Preheat oven to 350F (175C).
2. Combine flour, salt and pepper in a plastic bag. Add beef; shake to coat. Melt 3 tablespoons butter or margarine in a large skillet over medium heat. Add seasoned beef; sauté until browned. With tongs, transfer browned beef to a shallow baking pan large enough to hold beef in a single layer.
3. Add remaining butter or margarine to skillet. Add onion; sauté until soft. Add mushrooms; cook 1 minute.
4. Add stock to skillet; bring to a boil. Season mushroom mixture with salt and pepper; pour over beef.
5. Cover baking pan with foil or lid. Bake in preheated oven 20 minutes.
6. Add potatoes to casserole. Cover and bake 40 minutes longer or until potatoes and steak are tender.
7. To serve, use a slotted spoon to place beef and potatoes in a deep serving dish. Add half and half to cooking liquid left in baking pan. Heat liquid and half and half to make sauce; do not boil. Pour over steak. Garnish with parsley and carrot. Makes 4 servings.

Spicy Sweet & Sour Beef

1/4 teaspoon freshly grated nutmeg
1/2 teaspoon ground cinnamon
1 (3-1/2- to 4-lb.) beef-chuck cross-rib pot roast, boneless
8 whole cloves
1 large onion, sliced
1 garlic clove, crushed
2/3 cup packed brown sugar
1/2 cup wine vinegar
Salt
Freshly ground pepper
1/2 cup water
1 tablespoon cornstarch
Water

To garnish:
Cooked sliced zucchini, if desired

---

1. Combine nutmeg and cinnamon; rub over beef.
2. In a large glass or ceramic dish, combine cloves, onion, garlic, sugar, vinegar, salt, pepper and water; stir until sugar is dissolved. Add seasoned roast; turn to coat.
3. Cover and refrigerate 24 hours, turning several times.
4. To cook roast, preheat oven to 350F (175C). Place marinated roast and marinade in a large roasting pan. Cover pan with foil or lid. Bake in preheated oven 2-1/2 to 3 hours or until tender, basting once.
5. Place roast on a serving dish; cover with foil to keep warm. Discard cloves from cooking liquid.
6. Spoon fat from cooking liquid. In a small bowl, blend cornstarch with a little cold water; stir cornstarch mixture into cooking liquid in roasting pan. Bring to a boil, stirring constantly. Boil sauce rapidly 5 minutes or until slightly thickened and reduced in volume. Season sauce with salt and pepper. Pour some sauce over roast. Serve remaining sauce separately. Arrange zucchini around roast, if desired. Makes 8 to 10 servings.

Chicken Cacciatore

3 tablespoons vegetable oil
1 (2-1/2- to 3-lb.) broiler-fryer chicken, cut up or 4 chicken quarters
2 large onions, sliced
2 garlic cloves, crushed
1 (16-oz.) can tomatoes
2 tablespoons chopped fresh parsley or 1 tablespoon dried leaf parsley
1 teaspoon dried leaf basil
1/2 cup red wine
1 tablespoon tomato paste, if desired
Salt
Freshly ground pepper

This is a traditional Italian recipe in which chicken is flavored with garlic, tomatoes and basil.

1. Heat oil in a large skillet over medium heat. Add chicken; sauté until browned. Remove chicken from skillet; set aside.
2. Add onions and garlic to fat remaining in skillet; sauté until golden brown. Add tomatoes with their juice, parsley, basil, wine and tomato paste, if desired. Bring tomato mixture to a boil; season with salt and pepper.
3. Return chicken to skillet. Cover and simmer over low heat 1 hour or until tender.
4. Serve with favorite pasta. Makes 4 servings.

Boston Baked Beans

12 oz. pea beans or small white beans, soaked
4 oz. salt pork, cut into 1-inch pieces
2 large onions, sliced
1 teaspoon salt
1-1/2 teaspoons dry mustard
Freshly ground pepper
2 tablespoons molasses
2 tablespoons wine vinegar
8 whole cloves, tied in cheesecloth bag
1 tablespoon tomato paste or 2 tablespoons ketchup
Water

Boston Baked Beans is an American favorite. To do the recipe justice, long slow cooking is essential. First, wash beans thoroughly in cold water, then soak overnight in fresh cold water. A fast method of soaking is to boil un-soaked beans in fresh cold water in a large saucepan 2 minutes. Remove saucepan from heat; cover pan and let beans stand in cooking water 1 hour. Drain beans; cook according to recipe.

1. Preheat oven to 300F (150C).
2. Drain beans; place in a large heavy ovenproof casserole or bean pot.
3. Add salt pork to beans with onions, salt, mustard, pepper, molasses, vinegar, cloves and tomato paste or ketchup; stir well.
4. Add enough cold water to casserole to barely cover ingredients; cover casserole tightly with foil or lid. If lid is not a good fit, cover first with foil. Bake in preheated oven 5 to 6 hours or until beans are tender.
5. Check beans occasionally; add enough boiling water to keep beans covered during baking.
6. Discard bag containing cloves before serving. Makes 4 to 6 servings.

Oven Beef Stew

1-1/2 lbs. beef stew cubes
1 cup red wine
Grated peel of 1/2 orange
Juice of 1 orange
1 garlic clove, crushed
1 bay leaf
2 tablespoons vegetable oil
1 onion, chopped
1 tablespoon all-purpose flour
1 cup beef stock
Salt
Freshly ground pepper
2 to 3 tablespoons brandy
4 oz. button mushrooms, sliced

To garnish:
Baked puff-pastry triangles
Chopped fresh parsley

---

1. In glass or ceramic bowl combine beef, wine, orange peel and juice, garlic and bay leaf. Cover and refrigerate 3 hours.
2. Preheat oven to 325F (165C).
3. Drain beef, reserving marinade. Heat oil in a large skillet over medium heat. Add beef cubes; sauté until well browned. With a slotted spoon, transfer browned beef to a 3-quart casserole.
4. Add onion to fat remaining in skillet; sauté until lightly browned. Stir flour into onion; cook 1 minute. Add marinade and stock to skillet; bring to a boil. Season sauce with salt and pepper.
5. Warm brandy; pour over browned beef. Ignite brandy carefully. When flame dies, pour sauce over beef.
6. Cover casserole with foil or lid. Bake in preheated oven 2 hours.
7. Add mushrooms to hot casserole; bake an additional 30 to 45 minutes or until beef is tender. Discard bay leaf.
8. Garnish with pastry triangles and parsley. Makes 4 servings.

Variation
For a different flavor, add 1 cup prunes or 1 sliced red bell pepper 1 hour before end of cooking time. For an interesting flavor, add 2 tablespoons chopped fresh herbs or 1 tablespoon dried Italian seasoning or any individual herb, such as oregano, marjoram or basil.

Mexican Steak

1-1/2 lbs. beef-round steak, cut into 4 serving pieces
Salt
Freshly ground black pepper
3 tablespoons vegetable oil
1 onion, sliced
2 tablespoons all-purpose flour
1/2 teaspoon chili powder
1/4 teaspoon ground cumin
About 1-1/2 cups chicken stock
Hot-pepper sauce, if desired
1/2 cup canned whole-kernel corn, drained
1 avocado
Water
1 tablespoon lemon juice

1. Season beef with salt and pepper.
2. Heat oil in a large skillet over medium heat. Add beef; sauté until browned. With tongs, remove beef; set aside.
3. Add onion to fat remaining in skillet; sauté until soft. Stir in flour, chili powder and cumin; cook 2 minutes, stirring constantly. Stir in 1-1/2 cups stock; cook until slightly thickened. Season with salt, pepper and hot sauce, if desired.
4. Stir in corn. Return beef to skillet. Cover skillet; simmer 45 to 50 minutes or until beef is tender. Add additional stock, if necessary.
5. Peel and slice avocado. Place sliced avocado in a small bowl of water and lemon juice; set aside.
6. Arrange beef on a warmed serving dish. Add avocado to sauce; heat until warmed. Immediately spoon avocado and sauce over beef. Makes 4 servings.

Left to right: Mexican Steak, Oven Beef Stew

New England Chicken

1 (2-1/2- to 3-lb.) broiler-fryer chicken, cut up
Salt
Freshly ground white pepper
1/4 cup butter or margarine
1/2 cup milk
1/4 teaspoon dried rubbed sage
1/4 teaspoon dried leaf basil
1 (8-oz.) can oysters, drained
1/2 cup whipping cream
2 teaspoons cornstarch

To garnish:
Watercress

In New England, oysters are very plentiful in the fall. In other areas, fresh oysters may not be so easy to come by and may be expensive; however, canned oysters make a good alternative. If fresh oysters are available, use 1 cup shucked, drained fresh oysters.

1. Preheat oven to 350F (175C).
2. Sprinkle chicken with salt and pepper. Melt butter or margarine in a large skillet over medium heat. Add chicken pieces; sauté until golden. Transfer to a 3-quart casserole.
3. Pour milk over chicken; sprinkle with sage and basil.
4. Cover casserole with foil or lid; bake in preheated oven 1 hour.
5. Add oysters to casserole. In a small bowl, blend cream with cornstarch; stir cornstarch mixture into hot casserole. Recover casserole; bake 15 to 20 minutes longer or until chicken is tender.
6. Arrange chicken on serving platter. Spoon sauce and oysters over chicken; garnish with watercress. Makes 4 servings.

Australian Beef Curry

3 tablespoons vegetable oil
2 lbs. beef stew cubes
2 large onions, sliced
2 large cooking apples, peeled, cored, sliced
1 tablespoon curry powder
1 tablespoon all-purpose flour
1 cup beef stock
1 (16-oz.) can tomatoes
2/3 cup raisins
2 tablespoons wine vinegar
Salt

To garnish:
Hard-cooked eggs, sliced
Parsley sprigs
Homemade potato chips

Clockwise from left: New England Chicken, Greek-Style Oxtails, Australian Beef Curry.

1. Preheat oven to 350F (175C).
2. Heat oil in a large skillet over medium heat. Add beef cubes; sauté until brown. With a slotted spoon, transfer browned beef to a 3-quart casserole.
3. Add onions to fat remaining in skillet; sauté until soft. Add apples to skillet; cook 2 to 3 minutes.
4. Sprinkle curry powder and flour over onions and apples, stirring constantly. Stir in stock; bring to a boil.
5. Add tomatoes, raisins, vinegar and salt; spoon sauce over beef. Cover casserole with foil or lid.
6. Bake in preheated oven 2 hours. Stir casserole; bake an additional 15 to 30 minutes or until meat is tender.
7. To serve, spoon into a large serving dish. Garnish with hard-cooked egg, parsley and potato chips. Makes 4 to 5 servings.

Greek-Style Oxtails

2 oxtails, cut up
3 tablespoons olive oil
1/4 cup brandy
1 onion, sliced
1 garlic clove, crushed
1 cup dry white wine
2 bay leaves
Grated peel of 1/2 orange
Juice of 1 orange
Salt
Freshly ground pepper
2 cups beef stock
8 oz. pitted black olives
1/4 cup butter or margarine
1/3 cup all-purpose flour

---

1. Preheat oven to 325F (165C).
2. Trim oxtails of excess fat. Heat oil in a large skillet over medium heat. Add oxtails; sauté until browned. Transfer to a 3-quart casserole.
3. Warm brandy in a small saucepan; pour warmed brandy over oxtails. Ignite carefully; shake until flame dies.
4. Add onion and garlic to fat remaining in skillet; sauté until golden brown. Add wine to skillet; bring to a boil.
5. Pour wine mixture over oxtails; add bay leaves and orange peel and juice. Season with salt and pepper.
6. Bring stock to a boil in same skillet; add enough boiling stock to barely cover oxtails.
7. Cover casserole; bake in preheated oven 3 hours.
8. Pour cooking liquid into a medium bowl. Discard bay leaves. Refrigerate oxtails and cooking liquid overnight.
9. Next day, lift layer of fat from cooking liquid; discard fat. Bring cooking liquid back to a boil in a medium saucepan; pour over oxtails. Add olives. Cover casserole; bake 1 to 1-1/2 hours longer or until oxtails are tender.
10. Cream butter and flour together to make a beurre manié.
11. Strain cooking liquid into a medium saucepan. Whisk beurre manié into cooking liquid over medium heat, a little at a time, until thickened; bring back to a boil. Season with salt and pepper.
12. Arrange oxtails and olives in a serving bowl; pour sauce over oxtails. Makes 6 to 8 servings.

Oxtail has a high proportion of bone and is generally rather fatty. The flavor is excellent; however, it needs long, slow cooking. It is thus ideal for casseroles and stews. If oxtails are frozen, they should be thawed before cooking. If cutting your own, use a sharp knife and meat cleaver or saw. Cut into pieces about 2 inches thick; then trim off any excess fat.

Index